MOTHS - PESTS OF POTATO, MAIZE AND SUGAR BEET

Edited by **Farzana Khan Perveen**

Moths - Pests of Potato, Maize and Sugar Beet

http://dx.doi.org/10.5772/intechopen.73423

Edited by Farzana Khan Perveen

Contributors

Renata Bažok, Darija Lemić, Zrinka Drmić, Maja Čačija, Martina Mrganić, Helena Virić Gašparić, Silvia Rondon, Yulin Gao, Nisreen Alsaoud, Dommar Nammour, Ali Yassin Ali, Farzana Khan Perveen

Notice

Statements and opinions expressed in the chapters are these of the individual contributors and not necessarily those of the editors or publisher. No responsibility is accepted for the accuracy of information contained in the published chapters. The publisher assumes no responsibility for any damage or injury to persons or property arising out of the use of any materials, instructions, methods or ideas contained in the book.

First published in London, United Kingdom, 2018 by IntechOpen

IntechOpen is the global imprint of INTECHOPEN LIMITED, registered in England and Wales, registration number: 11086078, The Shard, 25th floor, 32 London Bridge Street
London, SE19SG – United Kingdom
Printed in Croatia

British Library Cataloguing-in-Publication Data
A catalogue record for this book is available from the British Library

Additional hard copies can be obtained from orders@intechopen.com

Moths - Pests of Potato, Maize and Sugar Beet, Edited by Farzana Khan Perveen
p. cm.
Print ISBN 978-1-78984-704-8
Online ISBN 978-1-78984-705-5

We are IntechOpen,
the world's leading publisher of
Open Access books
Built by scientists, for scientists

3,900+
Open access books available

116,000+
International authors and editors

120M+
Downloads

Our authors are among the

151
Countries delivered to

Top 1%
most cited scientists

12.2%
Contributors from top 500 universities

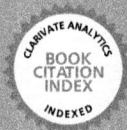

Interested in publishing with us?
Contact book.department@intechopen.com

Numbers displayed above are based on latest data collected.
For more information visit www.intechopen.com

Meet the editor

Dr. Farzana Khan Perveen (FLS; Gold-Medallist) obtained her BSc (Hons) and MSc (Zoology: Entomology) from the University of Karachi, MAS (Monbush-Scolar; Agriculture: Agronomy) from Nagoya University, Japan, and PhD (Research and Courseworks from Nagoya University; Toxicology) degree from the University of Karachi. She is Founder/Chairperson of the Department of Zoology (DOZ) and Ex-Controller of Examinations at Shaheed Benazir Bhutto University (SBBU) and Ex-Founder/Ex-Chairperson of DOZ, Hazara University and Kohat University of Science and Technology. She is author of 150 high-impact research papers, 135 abstracts, 40 authored books, 9 book chapters, and 8 edited books, and has supervised BS(4), MSc(75), MPhil(50), and PhD(1) students. She has organized and participated in numerous international and national conferences and received multiple awards and fellowships. She is a member of research societies, editorial boards of journals, the World Commission on Protected Areas, and the International Union for Conservation of Nature. Her fields of interest are entomology, toxicology, forensic entomology, and zoology.

Contents

Preface

The main purpose of this book is to present comprehensive and concise knowledge of the recent advancement on moths as pests of potato, maize, and sugar beet. Pests are destroying our crops worldwide. Control of pests has become a major issue and a crucial factor for future technological progress that must meet certain requirements to secure our crops. Overall the information compiled in this book will bring in-depth knowledge and recent advancement in the research.

Chapter 1 is an introductory chapter about moths. Perveen and Khan describe the history of moths, the difference with butterflies and skippers, classification, camouflage, navigation, attraction to light, and migration. Moths are useful as bio-indicators, pollinators, dispersal of seeds, food for other animals, nutrient recyclers, soil formers, and producers of useful products like silk threads. They are harmful as agricultural and stored-grain pests. They can be controlled biologically by predators and parasites, especially with *Bacillus thuringiensis*. They are also controlled by different types of pesticides such as farmers' pesticide, chemical pesticides, and plant pesticides, e.g., neem (*Azadirachta indica* Juss).

Chapter 2, "The Journey of the Potato Tuberworm Around the World," by Rondon and Gao reports that potato, *Solanum tuberosum* L. (Solanales: Solanaceae), production is challenged by many factors, including pests and diseases. Among insect pests, *Phthorimaea operculella* Zeller (Lepidoptera: Gelechiidae), known as the potato tuber worm/moth, is considered one of the most important potato pests worldwide. *Phthorimaea operculella* is a cosmopolitan pest of Solanaceous crops, including *S. tuberosum*, tomato (*Solanum lycopersicum* L.), and other important row crops. Adult moths oviposit in leaves, stems, and tubers; immature moths mine leaves causing foliar damage, but most importantly they burrow into tubers rendering them unmarketable. Currently, pest management practices are effective in controlling *P. operculella* but the effectiveness depends on many factors. Based on *P. operculella* biology, ecology, life cycle, distribution, seasonal dynamics, including its relationship with the potato crop, origins of potato crops, host range, and control methods, pest management practices can keep the pest under control. The effectiveness of control methods will depend on the response time to pest infestation, resources available, and also pest management practitioner experience. This chapter includes up-to-date information related to *P. operculella*.

Chapter 3, "Susceptibility of Egg Stage of Potato Tuber Moth, *Phthorimaea operculella* to Native Isolates of *Beauveria bassiana*," by Alsaoud et al. states that *P. operculella* females lay their eggs on the leaves and non-covered tubers near to the eyes (buds). Larvae dig tunnels during feeding, which causes damage to almost 100% of the cultivated and stored potato. Therefore, this moth must be controlled in the field and in the store. There are many ways to control this pest, starting with synthetic organic pesticides, natural origin insecticides such as botanical extracts, and by using genetically modified plants. Natural parasitic enemies are also successfully used such as wasps from Braconidae, in addition to insects predators from Coccinellidae, Chrysopidae, and Formicidae, and parasitic nematodes such as *Steinernema carpocapsae*, *Steinernema feltiae*, *Steinernema glaseri*, and *Heterorhabditis bacteriophora*, which are used successfully too. In the last decade, biological origin insecticides such as entomopathogenic viruses from the group baculovirus have been used, as well as entomopathogenic fungi like *Beauveria bassiana* (Hypocereales: Clavicipitaceae). In this chapter the pathogenicity of three native isolates of entomopathogenic fungus *B. bassiana* are studied in different concentrations on eggs of *P. operculella*. Findings indicate that eggs of *P. operculella* seem sensible to local isolates of *B. bassiana* in varying degrees.

Chapter 4 is entitled "Moths of Economic Importance in the Maize and Sugar Beet Production." According to Bažok et al., maize, *Zea mays* L. (Poales: Poaceae), and sugar beet, *Beta vulgaris* L. (Caryophyllales: Amaranthacea), production is often threatened by various pests, causing high yield losses. Economically, the most important maize pest is the European corn borer (ECB), *Ostrinia nubilalis* (Hübner), while the sugar beet moth, *Scrobipalpa ocellatella* (Boyd), and noctuid moths cause serious damage to *B. vulgaris*. This chapter highlights an introduction to several case studies representing long-term field research results on these pests. Depending on the pest, each study investigates the population level, dynamics of emergence or flight, damage levels, and possibilities of forecasting on different localities in Croatia. The results could be of great importance in the management of these pests. *Ostrinia nubilalis* management depends mainly on timely conducted control, but the damage level also depends on *Z. mays* hybrid and climatic conditions of the investigated area. Damage caused by *S. ocellatella* depends on the population level and on the locality's specific climate in a particular year. Scrobipalpa *ocellatella* population and flight dynamics can be monitored by using pheromones; however, pheromone application in forecasting and control has been shown to be disputable. Noctuid moths feed on *B. vulgaris* foliage, causing high damage, especially on young plants. The damage level depends on the climatic conditions of the research area, and visual inspections of caterpillars are necessary for forecasting and control decisions. The results of the investigation could be of great importance to the management of investigated pests, ECB, and moths (sugar beet moth and noctuid moths) on sugar beet. Results confirm that the damage to ECB is determined by weather conditions rather than by FAO maturity group. Noctuid moth damage to sugar beet leaves, determined by visual plant inspections, showed that the damage depends on climatic conditions of the location and decreases in very warm and dry conditions.

This book aims to provide readers with a comprehensive overview of moths as pests of potato, maize, and sugar beet and will focus on the most important research-oriented evidence of various advantageous aspects for parasitologists, entomologists, researchers, scientists, students, growers, field-men, producers and others that face the challenges imposed by these pests.

Dr. Farzana Khan Perveen (Gold-Medalist and FLS)
Founder Chairperson and Professor
Department of Zoology
Ex-Controller Examination
Shaheed Benazir Bhutto University (SBBU)
Main Campus, Sheringal, Dir Upper
Khyber Pakhtunkhwa, Pakistan

Introduction

Chapter 1

Introductory Chapter: Moths

Farzana Khan Perveen and Anzela Khan

Additional information is available at the end of the chapter

http://dx.doi.org/10.5772/intechopen.79639

1. Moths

The moths (Insecta: Lepidoptera) are the group of organisms allied to butterflies, having two pairs of wide wings shielded with microscopic scales. They are usually brightly colored and held flat at sitting posture. The word moths are derived from Scandinavian word mott, for maggot, perhaps a reference to the caterpillars of moths. Furthermore, about 165,000 species of moths, including micro- and macro-moths are found worldwide, many of which are yet to be described (**Table 1**) [1–3].

Kingdom: Animalia

  Subkingdom: Invertebrata

    Super-Division: Eumetazoa

      Division: Bilateria

        Subdivision: Ecdysozoa

          Superphylum: Tactopoda

            Phylum: Arthropoda Von Siebold, 1848

              Subphylum: Atelocerata

                Superclass: Hexapoda

                  Class: Insecta

                    Infraclass: Neoptera

                      Subclass: Pterygota

                        Superorder: Endopterygota

                          Unranked: Amphiesmenoptera

Unranked: Holometabola

Order: Lepidoptera Linnaeus, 1758

Examples:

- Micro-moths
- Macro-moths

Table 1. Taxonomic rank of moths [1].

1.1. History

Moths evolved long before butterflies, fossils have been found in Germany may be 200 million years old in the early Jurassic Period. Both types of lepidoptera (butterflies and moths), both adults and larvae are thought to have evolved along with flowering plants, mainly. In moths, the micro-lepidoptera tends to be more primitive in evolutionary terms than macro-lepidoptera [4]. Their fossils, some preserved in amber and some in very fine sediments. The earliest described lepidopteran taxon is *Archaeolepis mane*, a primitive moth-like species from the Jurassic, dated back to around 190 million years ago, and known only from three wings found in Dorset, Britain. The wings show scales with parallel grooves under a scanning electron microscope (SEM) and a characteristic wing venation pattern shared with caddis flies (Amphiesmenoptera: Trichoptera: ca. 14,500 described species) [5, 6].

1.2. Difference between moths with butterflies and skippers

Mostly, moths are dull-colored insects, with fat, hairy bodies, that fly at night, however, butterflies are brightly colored, delicate insects that fly during the day. Skippers are group of true butterflies but they are day flyers with fat fury bodies like moths [7]. In exception, many day-flying moths and a few butterflies and skippers fly in the early evening [8]. Moreover, a moth's antennae are long feathery (plumose). Further, butterflies have long thin antenna with clubbed tips. Furthermore, skippers have long thin antenna with clubs tapering to pointed hooks on the tip. In exception, several families of moths have antenna with clubs (Family: Castniidae), for example, the golden sun moth, *Synemon plana* Walker, 1854 [9]. However, moths fore- and hind-wings are held together with a structure called a frenulum. Moreover, butterflies and skippers wings are not joined. Further, in Australia with exceptions has only skipper in the world with a frenulum, for example, the regent skipper, *Euschemon rafflesia* (Macleay, 1827) (Family: Hesperiidae). It is only member of its genus, *Euschemon*, and Subfamily: Euschemoninae [10]. Moreover, many moths do not have a frenulum [11, 12]. Further, moths hold wings flat when resting. Furthermore, butterflies hold wings together above body. However, skippers' front-wings are held at a different angle to the back wings. In exception, many butterflies also rest with the wings flat [13]. Moreover, moth caterpillars spin a cocoon made of silk, around their body and pupate inside. Further, butterflies and skippers spin a pad of silk onto a stem or leaf then hang on the pad and form the pupae. In exception, many moths do not spin a cocoon and many butterflies and skippers spin a silken shelter attached with leaves (**Table 1** and **Figure 1**) [14].

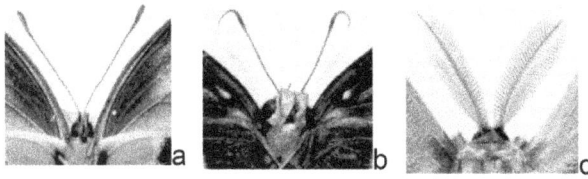

Figure 1. Types of antenna: (a) butterfly antenna; (b) skipper antenna; (c) moth antenna [10].

1.3. Navigation

Moths are navigated for their movement especially for migration. As one study of the moth heart and barbs showed that many moths may use the earth's magnetic field to navigate. The migratory behavior of the silver-Y, *Autographa gamma* (Linnaeus, 1758) (Family: Noctuidae) showed that even at high altitudes, the species can correct their direction, if their direction may change by winds. However, they to prefer fly with the direction of wind. If the wind is favorable to their direction, then it is easy for them to navigate during flying. Moths exhibit a tendency to circle artificial lights repeatedly. This suggests that they use a technique of celestial navigation called transverse orientation. By maintaining a constant angular relationship to a bright celestial light, such as the moon light, they can fly in a straight line. Celestial objects are so far away, even after traveling great distances, the change in angle between the moth and the light source is negligible. Further, the moon is always in the upper part of the visual field. When a moth encounters a much closer artificial light and uses it for navigation, the angle changes noticeably after only a short distance. The idea that moths may be impaired with a visual-distortion is called a Mach-band by Henry Hsiao in 1972. He stated that they fly towards the darkest part of the sky in pursuit of safety, thus they are inclined to circle ambient objects in the Mach-band region [15].

1.4. Attraction to light

There are many possible explanations to attract moths towards lights, but the most common theory is that many moths use the moon to navigate at night. By keeping the moon in a particular position, the moth can fly in a straight line and in the direction it wants. Unfortunately, they confuse bright lights for the moon and when they get close to the light. They cannot navigate properly and end up flying round and round in decreasing circles until they reach the source of the light and are burnt due to high intensity of heat [16].

1.5. Migration

Moths migrate in order to avoid antagonistic environmental conditions, like cold weather, starvation, drought and extremely hot weather. The short distances migrations relatively common among them. Gradually they survive and may lay eggs here after their arrival, but after hatching, their offspring do not survive in winter. One spectacular migrant commonly seen every year, is the wonderful hummingbird hawk moth, *Macroglossum stellatarum* (Linnaeus, 1758) (Bombycoidea: Sphingidae). Most of the migrants frequently migrate in hotter summers and when there are southerly winds. Other distinguished migrants, which seen every year are the African death head hawk moth, *Acherontia atropos* (Linnaeus, 1758); Greater death head hawk moth, *Acherontia lachesis* (Fabricius, 1798); lesser death head hawk moth, *Acherontia styx* Westwood, 1847 and the convolvulus hawk moth, *Agrius convolvuli* (Linnaeus, 1758) (Sphingidae: Sphiginae). Some species, like the crimson specked, *Utetheisa pulchella* (Linnaeus, 1758) (Noctuoidea: Erebidae) only occur in some years, but may sometimes arrive in large numbers. Perhaps, the most exotic looking migrant is the oleander hawk moth, *Daphnis nerii* (Linnaeus, 1758), which only arrives in some years and even in very low numbers. Another very communal migrant is the silver-Y-moth, *Autographa gamma* (Linnaeus, 1758) (Noctuoidea: Noctuidae) recognized by clear metallic Y-shape on each forewing [17].

1.6. Beneficial moths

Many moth species are beneficial for mankind as well as their ecosystem, the examples are:

1.6.1. Bioindicator

Moths are being affected by climate change. Species have always evolved to adapt to changing conditions. The problem with man-made climate change is that it is happening so quickly that moths may not be able to evolve and adapt fast enough. There have already been some winners and some losers as a result of climate change. One moth which has suffered is the beautiful garden tiger moth, *Arctia caja* (Linnaeus, 1758) (Noctuoidea: Erebidae). Sadly, this species is predicted to decline even further. The scarlet tiger moth, *Callimorpha dominula* (Linnaeus, 1758) (Noctuoidea: Erebidae) is found in many habitats, including gardens, and flies during the daytime in June and July. It particularly likes damp places and is often associated Russian comfrey, *Symphytum uplandicum* Linnaeus, 1753 (Boraginaceae: Lamiales) a favorite food of the caterpillars. The lime hawk moth, *Mimas tiliae* (Linnaeus, 1758) (Bombycoidea: Sphingidae) is another example for the same (**Figure 2**) [18].

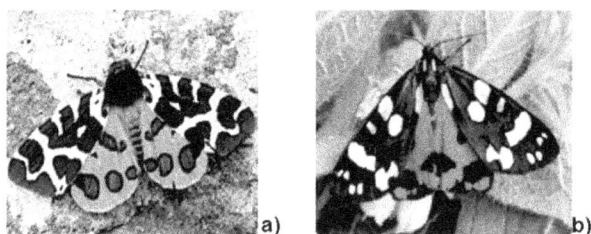

Figure 2. Insects as bioindicator: (a) the garden tiger moth, *Arctia caja* (Linnaeus, 1758) (Noctuoidea: Erebidae) and (b): the scarlet tiger moth, *Callimorpha dominula* (Linnaeus, 1758) (Noctuoidea: Erebidae) [18].

1.6.2. Useful products

Silk production (Sericulture) has a long history. Silk was discovered by Xilingji, wife of China's 3rd Emperor, Huangdi, in 2640 B.C. While making tea, Xilingji accidentally dropped a silkworm cocoon into a cup of hot water and found that the silk fiber could be loosened and unwound.

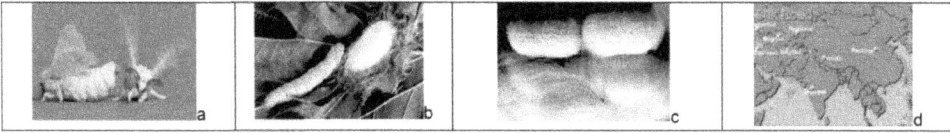

Figure 3. Sericulture: (a) adult silkworm, *Bombyx mori* (Linnaeus, 1758) (Bombycidae: Bobycinae; left: female; right: male); (b) *B. mori* Larva and silk cocoon; (c) silk strands and reeled from silk cocoons; (d) map of silk road [21, 22].

Fibers from several cocoons could be twisted together to make a thread that was strong enough to be woven into cloth. Thereafter, Xilingji discovered not only the means of raising silk worms, but also the manners of reeling silk and of employing it to make garments. Later sericulture spread throughout China, and silk became a precious commodity, highly sought after by other countries. Demand for this exotic fabric eventually created the lucrative trade route, the historically famous Silk Road named after its most important commodity [19]. The saturniids and bombycids yield silk of commercial value [20]. The silk moth, *Bombyx mori* (Linnaeus, 1758) (Bombycidae: Bobycinae) caterpillars are domesticated for silk. A number of wild moths such as *Bombyx mandarina* (Moore, 1872) (Bombycidae: Bobycinae) and *Antheraea Hübner*, 1819 species (Bombycidae: Sturniidae), besides others, provide commercially important silks (**Figure 3**) [21–23].

1.6.3. Food for other animals

Moths and their caterpillars are important food for many other species, including amphibians, small mammals, bats and many bird species. Moth caterpillars are especially important for feeding young chicks, including those of the most familiar garden birds such as *C. caeruleus* and great tit, *Parus major* Linnaeus, 1758 (Passeriformes: Paridae); robin, *Parus major* Linnaeus, 1758 (Passeriformes: Paridae); wren, *Troglodytes troglodytes* (Linnaeus, 1858) (Passeriformes: Troglodytidae) and blackbird, *Turdus merula* Linnaeus, 1758 (Passeriformes: Turdidae). Cuckoo, *Cuculus canorus* Linnaeus, 1758 (Cuculiformes: Cuculidae) specializes in eating hairy caterpillars, which most other birds avoid [24].

1.6.4. Pollinators and dispersal of seeds

Moths, including micro- and macro-moths are found worldwide are mostly feed on numbers of plant species. They travel flower to flower, plant to plant and place to place, therefore, they are responsible for pollination and dispersal of seeds for development of fruits and plants at different places, respectively [25].

1.6.5. Nutrients recyclers

Larvae of moths can cause damage to residential properties, like cloths, carpets, store grains just like termites may do. But they repay humans by performing a priceless service: helping us recycle decomposing dead materials. Decomposition may have an unpleasant ring to it but it is a fundamental process in a functioning ecosystem that is produced every year right on our own doorsteps. Larvae of moths wood eating are among the insect world's best decomposers organisms that digest dead matter [26].

1.6.6. Soil formation

Moths 93 species representing 10 families were recorded probing at soil and mud puddles. Observations of Gracillariidae and Lyonetiidae (97% males) are the first records at soil. Special mention is given those species of Geometridae and Notodontidae that pass large volumes of water through their gut as they drink from very wet substrates [26].

1.7. Camouflage

Moths show remarkable mimicry in different forms, which is still a challenge for evolution. Batesian mimicry is between palatable and non-palatable species, however, Mullerian mimicry, several equally unpleasantly tasting species share a color pattern and all species are mutually benefited, not only the mimic. They have significant economic importance.

The merveille-du-jour, *Griposia aprilina* (Linnaeus, 1758) (Family: Noctuidae) is a perfect match for lichen covered bark [27, 28]. The buff tip, *Phalera bucephala* (Linnaeus, 1758) (Family: Notodontidae) has gone one stage further and is not just the color of a twig, but the same shape too. A large group called the geometrids specializes in this disguise [29]. A few moths disguise themselves as distasteful, therefore, their predators will not even think of eating them. A moth looks just like a small bird dropping both in shape and color, the small bird lime moth, *Ponometia erastrioides* (Guenee, 1852) (Family: Noctuidae). Many moths use patterns that break up their outline, therefore, their moth shape is not recognized. A common garden moth or angle shades, *Phlogophora meticulosa* (Linnaeus, 1758) (Family: Noctuidae) combines several strategies (**Figure 4**) [30, 31].

Figure 4. Camouflage in moths: (a) merveille-du-jour, *Griposia aprilina* (Linnaeus, 1758) (Family: Noctuidae); (b): buff tip, *Phalera bucephala* (Linnaeus, 1758) (Family: Notodontidae); (c) Caterpillar (Family: Geometrids); (d) small bird dropping moth, *Ponometia erastrioides* (Guenee, 1852) (Family: Noctuidae); (e) angle shades, *Phlogophora meticulosa* (Linnaeus, 1758) (Family: Noctuidae); (f) lunar hornet clearwing moth, *Sesia apiformis* (Clerck, 1759) (Family: Sesiidae); (g) eyed hawk moth, *Smerinthus ocellatus* (Linnaeus, 1758) (Family: Sphingidae); (h) oleander hawk moth, *Daphnis nerii* (Linnaeus, 1758) (Family: Sphingidae); (i) rosy maple moth, *Dryocampa rubicunda* (Fabricius, 1793) (Family: Saturniidae); (j) copper underwing moth, *Amphipyra pyramidea* (Linnaeus, 1758) (Family: Noctuidae); (k) squeaking silk moth, *Rhodinia fugax* (Butler, 1877) (female) (Family: Saturniidae); (l) green silver lines (*Pseudoips prasinana*) (Linnaeus, 1758) (Family: Nolidae) [27–30].

1.8. Amazing moths

Moths use the tricks to avoid being eaten by their predators. Some members of the garden tiger moth, *Arctia caja* (Linnaeus, 1758) (Noctuoidea: Erebidae), which in the daytime use bright colors to warn predators that they taste bitter, and use squeaks in the dark to warn bats of their bad taste. The death's head hawk moth, *Acherontia atropos* (Linnaeus, 1758) (Sphingoidae: Sphingidae) makes squeaks, which apparently sound like those of a queen bee, fooling the worker bees into letting, it comes into their hive and eat their honey. The bee hawk moths, *Hemaris fuciformis* (Linnaeus, 1758) (Family: Sphingidae) have evolved to look just like bumble bees, *Bombus terrestris* (Linnaeus, 1758) (Hymenoptera: Apidae), predators think they can sting and will leave them alone. Female moths produce scents called pheromones to attract males, and the males use their antennae to pick up this scent as it wafts on the air. The male emperor moth, *Saturnia pavonia* (Linnaeus, 1758) (Family: Saturniidae) can often be seen following the scent towards females, and have been known to find them over distances of up to 5 miles. The caterpillar of the goat moth, *Cossus cossus* (Linnaeus, 1758) (Family: Cossidae) does not eat leaves but actually burrows into a tree trunk and eats the wood. Digesting wood is a slow process, therefore, the caterpillar takes 4 years to reach full size (**Figure 5**) [32, 33].

Figure 5. Amazing moth (a) banded tiger moth, *Apantesis vittata* (Fabricius, 1787) (Family: Arctiidae); (b) cinnabar moth, *Tyria jacobaeae* (Linnaeus, 1758) (Family: Eribidae); (c) bee hawk moths, *Hemaris fuciformis* (Linnaeus, 1758) (Family: Sphingidae); (d) emperor moth, *Saturnia pavonia* (Linnaeus, 1758) (Family: Saturniidae); (e) goat moth, *Cossus cossus* (Linnaeus, 1758) (Family: Cossidae); (f) lesser death head hawk moth, *Acherontia styx* Westwood, 1847 (Family: Sphingidae) [32].

1.9. Harmful moths: as pests

The larvae of many moth species are significant pests of agricultural crops and stored grains. They cause great losses to mankind. Some reports estimate that there have been over 80,000 caterpillars of several different taxa feeding on a single oak tree, *Quercus akoensis* Mull. (Fagales: Fagaceae). The major pest families are Tortricidae, Noctuidae and Pyralidae. Well-known species are the cloth moths, *Tineola bisselliella* (Hummel, 1823); *T. pellionella* Linnaeus, 1758, and carpet moth, *Trichophaga tapetzella* (Linnaeus, 1758) (Tineoidae: Tineidae), feeding on foodstuffs. They have also been found on bran, semolina, flour (e.g., wheatflour), biscuits, casein and insect specimens in museums [34]. The larvae of the Noctuidae, army worms, *Spodoptera frugiperda* (Smith, 1797) and corn earworm, *Helicoverpa* zea (Hübner, 1808)

(*Noctuoidea*: Noctuidae) can cause extensive damage to certain crops. The cotton boll worms, *Helicoverpa armigera* (Hübner, 1808) (*Noctuoidea*: Noctuidae) larvae are polyphagous. The variegated cutworms, *Peridroma saucia* (Hübner, 1808) (*Noctuoidea*: Noctuidae) are described as one of the most damaging pests to gardens. Throughout the world, the diamondback moth (DBM), *Plutella xylostella* L. (Lepidoptera: Plutellidae) is considered the main insect pest of brassica crops, particularly, the cabbage, *Brassica oleracea* or variants L.; white cabbage (*capitata* var. *alba* L.); kales crops, *red Russian kale, Brassica napus* L. subsp. *napus* var. *pabularia* (DC.) Alef.; broccoli, *Brassica oleracea* L. (cultivar group: Italica); and cauliflower, *Brassica oleracea* L. (Brassicales: Brassicaceae). It has been known to completely destroy *B. oleracea* (*capitata* var. *alba*) and *B. napus*. In Kenya, *P. xylostella* has also been found feeding on peas, *Pisum sativum* L. (Fabales: Fabaceae) [35]. Pesticides can affect other species than the species they are targeted to eliminate and damaging the natural ecosystem [36].

1.9.1. Biological control of moths

Biological control is relatively permanent, safe, economic and environmental friendly [37]. The crops management practices include protects and encourages natural enemies and increases their impact on pests for conservation as a biological control method [38, 39]. The parasitoid stingless wasp, *Trichogramma chilonis* (Ishii) (Hymenoptera: Trichogrammatidae) is an important egg parasitoid used for the control of the Mediterranean flour moth, *Ephestia kuehniella* (Zell, 1879) (Pyraloidea: Pyralidae); angoumois grain moth, *Sitotroga cereallela* (Olivier, 1789) (Gelechiioidea: Gelechiidae) and rice meal moth, *Corcyra cephalonica* (Stainton, 1866) (Pyraloidea: Pyralidae). *Sitotroga cerealella* originally proposed by Flanders [40, 41] is one of the most commonly used as fictitious host for rearing *Trichogramma sp.* The pyralid cactus moth, *Cactoblastis cactorum* (Berg, 1885) (Pyraloidea: Pyralidae) was introduced from Argentina-Australia, where it successfully suppressed millions of acres of prickly pear cactus, *Opuntia abjecta* Britton and Rose (Caryophyllales: Cactaceae: Opuntioideae). Another species of the Pyralidae, called the alligator weed stem borer, *Arcola malloi* (Pastrana, 1961) (Pyraloidea: Pyralidae) was used to control the aquatic plant known as the alligator weed, *Alternanthera philoxeroides* (Mart.) Griseb (Caryophyllales: Amaranthaceae) in conjunction with the alligator weed flea beetle, *Agasicles hygrophila* Selman and Vogt, 1971 (Galerucinae: Agasicles); in this case, two insects work in synergy and the weed rarely recovers [42].

1.9.2. Bacillus thuringiensis

Bacillus thuringiensis (Bt) var. *aizawai* and Bt var. kurstaki are very effective in controlling infestations of *P. xylostella* (Lepidoptera: Plutellidae). Bt var. *kurstaki* is widely used at a weekly interval and a rate of 0.5/ha. Bt kills the *P. xylostella* and does not harm beneficial insects [43].

1.9.3. Farmers' pesticide

Farmers produce their own homemade biopesticides by collecting diseased *P. xylostella* caterpillars (fat and white or yellowish or with fluffy mold on them), crushing and mixing them with water in a blender. Large tissue clumps are filtered out and the liquid is sprayed [44].

1.9.4. Neem, Azadirachta indica

The neem-based products, *Azadirachta indica* Juss 1830 (Sapindales: Meliaceae) give a good control of *P. xylostella* and are relatively harmless to natural enemies [45].

Author details

Farzana Khan Perveen[1]* and Anzela Khan[2]

*Address all correspondence to: farzana_san@hotmail.com

1 Department of Zoology, Shaheed Benazir Bhutto University (SBBU), Sheringal, Khyber Pakhtunkhwa (KP), Pakistan

2 Roots Millennium College (RMC), Islamabad, Pakistan

References

[1] Hegna TA, Legg DA, Møller OS, Van R, Lerosey AR. The correct authorship of the taxon name Arthropoda (PDF). Arthropod Systematics and Phylogeny. 2013;**71**(2):71-74

[2] McGAvin G. Entomología Esencial. Barcelona, España: Ariel Ciencia; 2002. pp. 1-350

[3] Ugarte A. Pequeña guía de Campo. In: Mariposas de Chile. Santiago, Chile: Ograma Impresores; 2015. pp. 1-250

[4] Kaplan M. Moths jam bat sonar, throw the predators off course. In: National Geographic News. Cambridge, United Kingdom (UK): Cambridge University Press; 2009. Online: http://news.nationalgeographic.com/news/2009/07/090717-moths-jam-bat-sonar.html [Accessed: March 23, 2018]

[5] Grimaldi D, Engel MS. Evolution of the Insects. Cambridge, United Kingdom (UK): Cambridge University Press; 2005. p. 136

[6] Samways MJ. Insect Diversity Conservation. Cambridge, United Kingdom (UK): Cambridge University Press; 2005. pp. 1-342

[7] Mallet J. Taxonomy of Lepidoptera: The Scale of the Problem. The Lepidoptera Taxome Project. London, United Kingdom (UK): University College; 2007. pp. 1-56

[8] Miller JC, Hammond PC. Lepidoptera of the Pacific Northwest Caterpillars and Adults. USA: Forest Health Technology Enterprise Team (FHTET), Technology Transfer, USAD; 2003. pp. 1-323

[9] Shield O. World number of butterflies. Journal of Lepidopterists'Society. 1989;**43**:178-183

[10] Covell CC Jr. A Field Guide to Moths. Boston, MA: Houghton Mifflin Co.; 1984. pp. 1-496

[11] Covell CC Jr. The butterflies and moths of Kentucky, an annotated checklist. Kentucky State Nature Preserves Commission. Scientific and Technical Series. 1999;**6**:1-220

[12] Sutton P, Sutton C. How to Spot Butterflies. Boston, USA: Houghton Mifflin Company; 1999. pp. 1-141

[13] Khan AG, Azim A, Nadeem MA, Ayaz M. The effect of formaldehyde treatment of solvent and mechanical extracted cottonseed meal on the performance, digestibility and nitrogen balance in lambs. Asian-Australasian Journal of Animal Sciences. 2000;**13**(6):785-790

[14] Haroon K, Perveen F. Distribution of butterflies (Family: Nymphalidae) in Union Council Koaz Bahram Dheri, Khyber Pakhtunkhwa, Pakistan. Social and Basic Sciences Research Review (SBSRR). 2015;**3**(1):52-57

[15] Chapman A, Jason W, Reynolds DR, Mouritsen HH, Jane K, Riley JR, Sivell D, Smith AD, Woiwod IP. Wind selection and drift compensation optimize migratory pathways in a high-flying moth. Current Biology. 2008;**18**(7):514-518. DOI: 10.1016/j.cub.2008.02.080. PMID 18394893

[16] Robert SB, Oliveira A, Evandro G, Riveros AJ. Experimental evidence for a magnetic sense in Neotropical migrating butterflies (Lepidoptera: Pieridae) (PDF). The British Journal of Animal Behaviour. 2005;**71**(1):183-191. DOI: 10.1016/j.anbehav.2005.04.013

[17] Smith NG, Janzen DH, editors. Urania Fulgens (*Calipato Verde, Green Urania*). Costa Rican Natural History. Chicago, USA: University of Chicago Press; 1983. pp. 1-816

[18] Saunders D, Hobbs R, Margukes C. Biological consequences of ecosystem fragmentation: A review. Conservation Biology. 1991;**5**:18-31

[19] Good LI, Kenoyer JM, Meadow R. New evidence for early silk in the Indus civilization. Archaeometry. 2009;**51**(3):457-466. DOI: 10.1111/j.1475-4754.2008.00454.x

[20] Perveen F, Khan A, Sikandar A. Characteristics of butterfly (*Lepidoptera*) fauna from Kabal, Swat, Pakistan. Journal of Entomology and Zoology Studies. 2014;**2**(1):56-69. Online: http://www.entomoljournal.com/vol2Issue1/21.1.html [Accessed: March 24, 2018]

[21] Cherry R. History of sericulture: Silk production; cultural entomology digest. Insect Facts and History. 2018;**5**:1-2. Online: http://www.insects.org/ced1/seric.html [Accessed: March 24, 2018]

[22] Wikipedia FE. Sericulture. 2018. Available from: http://en.wikipedia.org/wiki/Sericulture [Accessed: March 24, 2018]

[23] Vainker S. Chinese Silk: A Cultural History. New Jersey, USA: Rutgers University Press; 2004. pp. 1-20

[24] Resh VH, Ring TC. (1 July). Encyclopedia of Insects. 2nd ed. Cambridge, USA: USA Academic Press; 2009. pp. 1-2

[25] Perveen F, Ahmad A. Check list of butterfly fauna of Kohat, Khyber Pakhtunkhwa, Pakistan. Art. 2012;**1**(3):112-117. Online: http://www.iaees.org/publications/journals/arthropods/arthropods.asp; http://easletters.com/ [Accessed: March 24, 2018]

[26] Adler H. Soil- and puddle-visiting habits of moths l peter. Journal of the Lepidopterists' Society. 1982;**36**(3):161-173

[27] Ferguson DC, Chuck EH, Paul AO, Richard SP, Michael P, Jerry AP, Michael JS. Moths of North America. Jamestown, ND: Northern Prairie Wildlife Research Center; 1999. pp. 1-2. Online: http://www.npwrc.usgs.gov/resource/distr/lepid/moths/mothsusa.htm [Accessed: May 15, 2017]

[28] Wickler W. Mimicry in Plants and Animals. Oxford, United Kingdom (UK): McGraw-Hill; 1968. pp. 1-2

[29] Meyer A. Repeating patterns of mimicry. Publication of Library Science Biology. 2006; **4**(10):1-675

[30] Stevens M, Merilaita S. Animal Camouflage: Mechanisms and Function. Cambridge, United Kingdom (UK): Cambridge University Press; 2011. pp. 1-2

[31] Arjen E, van't H, Pascal C, Daniel JR, Carl JY, Jessica L, Michael A, Quail NH, Alistair CD, Ilik JS. The industrial melanism mutation in British peppered moths is a transposable element. Nature. 2016;**534**:102-105. DOI: 10.1038/nature17951. PMID: 27251284

[32] Hodges RW et al. Check List of the Lepidoptera of America North of Mexico. London, United Kingdom (UK): E. W. Classey Ltd; 1983. pp. 280-284

[33] Hans B. Moths with a taste for tears: Insects that live off the tears of mammals find the secretions a tasty and nutritious food. New Scientist. 1990;**38**:1-4

[34] Albert G. Eigenartige Geschmacksrichtungen bei Kleinschmetterlingsraupen [Strange Tastes Among Micromoth Caterpillars] (PDF) (in German). Cambridge, United Kingdom (UK): Cambridge University Press; Vol. 271942. pp. 105-109

[35] Khan ZR, Midega CAO, Wadhams LJ, Pickett JA, Mumuni A. Evaluation of Napier grass (*Pennisetum purpureum*) varieties for use as trap plants for the management of African stemborer (*Busseola fusca*) in a push–pull strategy. Entomologia Experimentalis et Applicata. 2007;**124**(2):201-211

[36] Forbes L. Lepidoptera of New York and Neighboring States. Part III. Geometridae, Sphingidae, Notodontidae, Lymantriidae. Cornell Agriculture Experimental Station Memory. Cornell, USA: Cornell University Press; Vol. 2741948. pp. 259-263

[37] Debach P. Biological Control by Natural Enemies. London, UK: Cambridge University Press; 1974. pp. 1-342

[38] Perveen F, Sultan R, Haque EU. Role of temperature and host (*Sitotroga cereallela* and *Corcyra cephalonica*) egg age on the quality production of *Trichogramma chilonis*. Art. 2012;**1**(4):144-150. Online: http://www.iaees.org/publications/journals/arthropods/asp [Accessed: March 24, 2018]

[39] Corrigan JE, Laing JE. Effects of the rearing host species and the host species attacked on performance by *Trichogramma minutum* Riley (Hymenoptera: Trichogrammatidae). Environmental Entomology. 1994;**23**:755-760

[40] Flanders SE. Mass production of egg parasites of genus *Trichogramma*. Hilgardia. 1930; **4**:465-501

[41] Singh SP, Singh J, Brar KS. Effect of temperature on different strains of *Trichogramma chilonis* Ishii. Insect Environment. 2002;**7**:181-182

[42] Perveen F, Sultan R. Effects of the host and parasitoid densities on the quality production of *Trichogramma chilonison* lepidopterous (*Sitotroga cereallela* and *Corcyra cephalonica*) eggs. Art. 2012;**1**(2):63-72. Online: http://www.iaees.org/publications/journal/arthropods/arthropods.asp; http://easletters.com/ [Accessed: March 24, 2018]

[43] Irshad M. Biological Control of Insects and Weeds in Pakistan. Islamabad, Pakistan: HEC Report; 2008. pp. 1-315

[44] Perveen F, Khan A. Comparison of integrated pest management techniques used to control the diamondback moth, *Plutella xylostella* L. on cauliflower, *Brassica oleracea* L. Global Journal of Agricultural Research (GJAR). 2013;**1**(1):8-22. Online: http://www.eajournals.org/journals/global-journal-of-agricultural-research-gjar/vol-1-issue-1-june-2013/comparison-of-integrated-pest-management-technique-used-to-control-the-diamondback-moth-plutella-xylostella-on-cauliflower-brassica-oleracea/ [Accessed: March 24, 2018]. Published by European Centre for Research Training and Development UK

[45] Rushtapakornchai W, Vattanatangum A, Saito T. Development and implementation of sticky trap for diamondback moth control in Thailand. In: Diamondback Moth and Other Crucifer Pests. Proceedings of the Second International Workshop. Talekar NS, editor. Asian Vegetable Research and Development Center; Shanhua, Taiwan; 1992. pp. 523-528. Online: www.avrdc.org [Accessed: March 24, 2018]

Moths as Pests

Chapter 2

The Journey of the Potato Tuberworm Around the World

Silvia I. Rondon and Yulin Gao

Additional information is available at the end of the chapter

http://dx.doi.org/10.5772/intechopen.81934

Abstract

Potato (*Solanum tuberosum* L.) production is challenged by many factors including pests and diseases. Among insect pests, *Phthorimaea operculella* Zeller (Lepidoptera: Gelechiidae), known as the potato tuber worm or potato tuber moth, is considered one of the most important potato pests worldwide. *Phthorimaea operculella* is a cosmopolitan pest of solanaceous crops including potato, tomato (*Solanum lycopersicum* L.), and other important row crops. Adults oviposit in leaves, stems, and tubers; immature stage mines leaves causing foliar damage, but most importantly, burrows into tubers rendering them unmarketable. Currently, pest management practices are effective in controlling *P. operculella*, but the effectiveness depends on many factors that will be discussed later in this chapter. Each section includes up-to-date information related to *P. operculella* biology, ecology, and control, including origins, host range, life cycle, distribution, seasonal dynamics, and control methods.

Keywords: *Phthorimaea operculella*, lepidoptera, gelechiidae, moth, pest management, solanaceous, IPM, tubermoth

1. Introduction

Potato (*Solanum tuberosum* L.) is the fourth major food crop around the world after rice (*Oryza sativa* L.), wheat (*Triticum* spp.), and maize (*Zea mays* L.). Potato is rich in vitamins, minerals, proteins, antioxidants, essential amino acids, and carbohydrates [1–4], and it is an important part of many cultures diet around the world. Indians from Peru were the first ones to cultivate potatoes around 8000–5000 BC; by the early 1500s, when the Spaniards arrived to South America, they brought first potato plants to Europe, and eventually, potatoes were genetically improved and grown, and nowadays, the crop is utilized worldwide [5].

Potato production is challenged by many factors including pests and diseases. Depending on where potato production occurred, aphids (*Macrosiphum euphorbiae* Thomas and *Myzus persicae* Sülzer), leafhoppers, potato psyllids (*Bactericera cockerelli* Šulc), beetles (*Leptinotarsa decemlineata* Say), and moths can have a tremendous effect on the crop. *Phthorimaea operculella* Zeller (Lepidoptera: Gelechiidae), also known as the potato tuber worm or potato tuber moth (**Figure 1**), is considered one of the most important potato pests worldwide [6–14]. *Phthorimaea operculella* is a cosmopolitan pest of solanaceous crops including potato, tomato (*Solanum lycopersicum* L.), and other important row crops [15–19]. Adults oviposit in leaves, stems, and tubers; the immature stage mines leaves, (**Figure 2**) but most importantly, larvae burrow into tubers rendering them unmarketable (**Figure 3**); unfortunately, outbreaks are still difficult to predict. Pest management practices are effective in controlling *P. operculella*

Figure 1. *Phthorimaea operculella* Zeller adult. Photo credit: Oregon State University, Extension and Experiment Station Communication (Ketchum).

Figure 2. *Phthorimaea operculella* Zeller larva. Photo credit: Oregon State University, Extension and Experiment Station Communication (Ketchum).

Figure 3. *Phthorimaea operculella* Zeller tuber damage. Oregon State University. Irrigated Agricultural Entomology Program (Rondon).

but the effectiveness depends on the response time to pest infestation, resources available, and pest management practitioner experience. This chapter includes a compilation of up-to-date information related to *P. operculella* biology, ecology, and control, including origins, host range, life cycle, distribution, seasonal dynamics, and control methods.

2. *Phthorimaea operculella* as an agricultural pest around the world

Phthorimaea operculella was first reported as pest-affecting tubers in South America in the early 1900s [20, 21]. Foliar damage does not usually result in significant yield losses [20]; however, reduce marketability and damage due to tuber infestation can be significant in nonrefrigerated storage conditions [22]. For instance, in the Middle East, *P. operculella* infestation can range between 1 and 65% [23, 24], while in India, *P. operculella* is responsible for about 1–13% and 70–100% infestation in the field and storage, respectively [25–29]. In Ethiopia, *P. operculella* is responsible for 9–42% yield loss [30]; Lagnaoui et al. [31] indicated that yield loss in storage could be up to 100% where no temperature and/or humidity control is possible.

2.1. Scientific nomenclature of the potato tubeworm

The taxonomic tree of *P. operculella* is illustrated in **Figure 4**. This insect belongs to the Phylum Arthopod, Class Insecta, Order Lepidoptera, Sub-Order Glossata, Super Family Gelechioidea, Family Gelechiidae, and Sub-Family Gelechiinae. *Phthorimaea operculella* was described in 1873 as *Bryotropha* and then *Gelechia operculella* [32]. The genus was revised in 1902 and 1931 and assigned to the genus *Phthorimaea* in 1964 by Meyrick [33], Povolny [34], and Povolny and Weissman [35]. In the old literature, *P. operculella* can be found as *Gnorimoschema operculella, Lita operculella, Lita solanella, P. solanella,* and *P. terrella*; more recently as *Scrobipalpa operculella, Scrobipalpus solanivora,* and *S. solanivora*; until finally recognized as *P. operculella* [36].

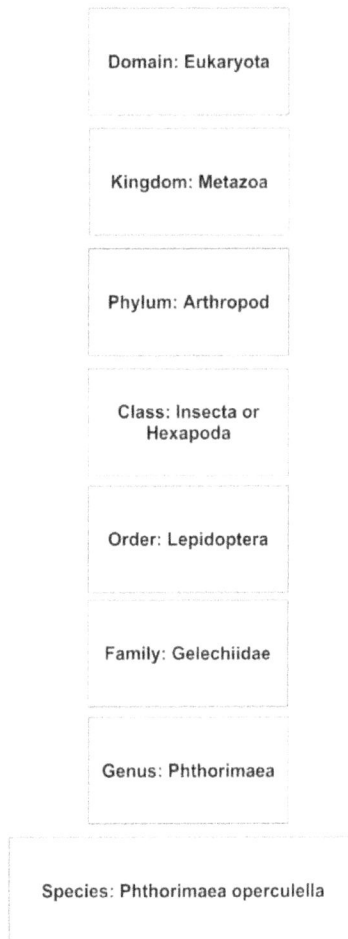

Figure 4. Taxonomical tree of *Phthorimaea operculella*. Photo credit. Oregon State University. Irrigated Agricultural Entomology Program (Rondon).

Although in the Entomological Society of America database, the common name of *P. operculella* is potato tuberworm [37], other recognized regional names are listed in **Table 1**. Currently, two other species of Gelechiidae moths are known as "tuber worms," *Tecia solanivora* (Povolny), restricted to Central and Northwest South America known as the Central American potato tuberworm or Guatemalan potato moth, and *Symmestrischema plaesiosema* (Turner), found in South America, Southeast Australia, and Philippines [19, 38]. This chapter will focus on the *Phthorimaea* species.

2.2. Distribution

Currently, *P. operculella* can be found in all potato production areas in tropical and subtropical countries in South, Central, and North America, Africa, Australia, and Asia [14, 40–42].

Language	Common names
English	Potato moth; potato tuber worm; stem end grub; tobacco leafminer; tobacco split worm; tobacco splitworm
Spanish	Gusano de la papa; gusano del tubérculo de la papa; minador común de la papa; minador de la hoja del tabaco; oruga barrenadora del tallo; palomilla de la patata; polilla de la papa; polilla de la patata
French	Teigne de la pomme de terre
German	Kartoffelmotte
Italian	Tignola della patata
Danish	Kartoffelmol
Dutch	Aardappel-knollenruspje
Hebrew	Ash habulbusin

Table 1. Common names of *Phthorimaea operculella* Zeller [36, 39].

Phthorimaea operculella is the single-most significant insect pest of potato (field and storage) in North Africa, Asia, and the Middle East [43, 44]. In the mid-1800s, *P. operculella* was reported in Tasmania, New Zealand, and Australia [45]. In South Central Asia, *P. operculella* was introduced in 1906 to Bombay, India, apparently from Italy [46]; but by the mid-1900s, *P. operculella* became widely distributed in all potato regions in India. In 1913, *P. operculella* was first reported in the USA [14, 47–50]; at present, *P. operculella* is present in most USA potato production regions [14, 51]. The first report in China took place in 1937, when Chen found *P. operculella* larvae in tobacco (*Nicotiana tabacum* L.) plants in the Liuzhou City in Guangxi province [52–54]. In the mid-1970s, *P. operculella* was introduced to Iraq [24, 55], and by the early 1980s, it was found in Russia [56]. From 2002, *P. operculella* has emerged as a problem in the Bologna providence in northern Italy [57]. To our knowledge, there are few studies describing the population structure of *P. operculella* around the world that could explain *P. operculella* distribution. In the USA, Medina and Rondon [58] suggested that geographical barriers such as the Appalachian Mountains in North America might act as a geographic barrier isolating *P. operculella* sub-populations. *Tuta absoluta* Meyrick (Lepidoptera: Gelechiidae), a close relative of *P. operculella*, is a serious pest of tomatoes in Europe, Africa, western Asia, South America, and Central America and can sometimes be taxonomically confused with *P. operculella* [59, 60].

2.3. Host range

Phthorimaea operculella is primarily a pest of potato but also can be found in other solanaceous crops and weeds (**Table 2**). There is only one report unconfirmed of *P. operculella* in sugar beet [61]. Das and Raman [16] reported alternate hosts representing 60 plant species, both cultivated and wild. Most of the hosts belong to the Solanaceae family, while others belong to the Scrophulariaceae, Boraginaceae, Rosaceae, Typhaceae, Compositae, Amaranthaceae, and Chenopodiaceae.

Phthorimaea operculella can be found in all crops and weeds listed above; however, field studies have shown that *P. operculella* only reproduce when feeding on potato, tomato, sugar

Host Common name	Scientific name	Family
Potato	*Solanum tuberosum*	Solanaceae
Eggplant, aubergine	*Solanum melongena* L.	Solanaceae
Bell pepper	*Solanum annuum* L.	Solanaceae
Tomato	*Solanum lycopersicum* L.	Solanaceae
Black nightshade	*Solanum nigrum* L.	Solanaceae
Silver leaf nightshade	*Solanum elaeagnifolium* Cav.	Solanaceae
Chili pepper	*Capsicum frutescens* L	Solanaceae
Tobacco	*Nicotiana tabacum* L	Solanaceae
Cape gooseberry	*Physalis peruviana* L.	Solanaceae
Field ground cherry	*Physalis mollis* D.	Solanaceae
Prickly nightshade	*Solanum torvum* Sw.	Solanaceae
Jimson weed	*Datura stramonium* L.	Solanaceae
Gooseberry	*Physalis angulate* L.	Solanaceae
Angel's tears	*Brugmansia suaveolens* Bersch	Solanaceae
	Fabina, Lycium, Hyoscyamis, Nicandra	Solanaceae
Sugarbeet	*Beta vulgaris var. saccharifera*	Amaranthaceae

Table 2. List of common hosts of *Phthorimaea operculella* Zeller [28, 61–66].

beet, and eggplant [16, 28, 47, 61, 67–71]. In the western USA states, *P. operculella* was found to feed and reproduce only in potatoes [14, 50]. Early behavior of first instars is critical for establishing in a suitable host plant [72]; thus, not surprisingly, food source availability and quality are critical in *P. operculella* establishment and success [14, 50]. A pattern in the diet of many groups of herbivorous insects is that related species tend to feed on related co-evolutionary plants; however, do not dismiss the adaptability of the insect taxa [73]. A great example of this statement is the co-evolution of the potato crops and the Colorado potato beetle, *Leptinotarsa decemlineata* Say [74]. All instars of *P. operculella* can potentially survive in volunteer potatoes or in the soil [50, 75–77] and have high adaptability to cold acclimation, cooling rate, and heat stress [78].

3. Biological and ecological aspects

3.1. Biology

Phthorimaea operculella has four life stages: adult, egg, larva, and pupa. Development, survival, and reproductive rates vary considerably in relationship to host quality and availability [79].

3.1.1. Adults

Adults are small moths (approximately 0.94 cm long) with a wingspan of approximately 1.27 cm. Forewings have 2–3 dark dots on males and an "X" on females [14, 29, 50, 80, 81] (**Figure 5**). Both pairs of wings have characteristic-fringed edges. Early literature considered that adults were poor fliers [9, 14, 15, 82]; however, recent studies have shown that they can fly for over 5 hours or up to 10 km nonstop [83]. Krambias [84] and Foley [83] indicated that *P. operculella* cannot fly at wind speeds in excess of about 5–6 m/s. Moths are active at temperatures between 14.4 and 15.5°C; at around 11.1˚C, they can crawl through soil cracks or burrow short distances through loose soil. In Oregon, *P. operculella* was observed searching for tubers at temperatures close to 5°C. [85]. Copulation can take place only 16–20 hours after adult emergence; the duration of copulation ranges between 85 and 200 minutes [28, 86, 87]. Adults can live for 1–2 weeks [77]. Adults are normally inactive during the day and oviposition occurs at night [68, 69, 88–90]. Adults do not oviposit in the soil close to tubers if potato foliage is available [14, 89, 91]. Eggs laid and their longevity are directly related to their nutrition [14, 90, 92, 93], and age of male appears to play an important role in the ability to mate [94]. Selection of plants for oviposition is determined by the physical nature of plant surface and by chemical factors that are detected only when females enter in contact with the host [61]. Meisner et al. [70] showed that oviposition is stimulated by ethanolic extracts and I-glutamic acid released from potato peels. Chemical cues are mainly responsible for host selection; olfactory detection of plant volatiles may elicit the female to find the best host for her offsprings [95].

3.1.2. Eggs

Eggs are ≤0.1 cm spherical, translucent when freshly laid turning white or yellowish to light brown after 1–2 hours. In the field, females lay their eggs on foliage, soil and plant debris, or

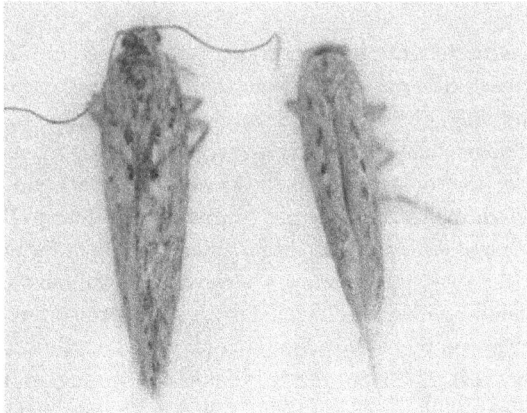

Figure 5. *Phthorimaea operculella* female (left) and male (right). Photo credit. Oregon State University. Irrigated Agricultural Entomology Program (Rondon).

exposed tubers [14, 50]; however, foliage was the preferred oviposition substrate [72]. There are discrepancies related to number of eggs per batch [24, 77, 96, 97]. For instance, Gubbaiah and Thontadarya [28] indicated that in the field, females laid eggs singly and rarely in groups of 3–5 eggs on either side of the leaf but close to the mid-rib. In the storage (~7.2°C), eggs were laid singly or in groups of 3–15 near the eye buds. Regarding distribution of eggs in plants, eggs can be widely distributed with greater numbers found around the base of the plants [89]; in confinement, *P. operculella* oviposits in groups close to eye buds [85]. Incubation period could range from 5 to 34 days [65], 4–5 days [98], 2.3 and 7.2 days at 33.3 and 20.9°C, respectively [24], 3–10 days [24, 63, 99, 100]. Attia and Mattar [68] reported 36°C as the upper critical temperature at which no eggs were laid.

3.1.3. Larvae

Larvae are usually cream to light brown reddish with a characteristic brown head. Mature larvae (0.94 cm long) may have a pink or greenish color; thorax has small black points and bristles on each segment [77]. No sexual dimorphism is observed until the third larval stage where initial sexual structures are visible; in the 4th larval stage, males are different from females where males have two elongated yellowish testes in the 5th and 6th abdominal segment [29]. Moregan and Crumb [101] reported 15–17 days for the larval period; Graft [20] and Trivedi and Rajagopal [65] reported 13–33 days, and Van der Goot [98] reported 14 days. Larvae feed on leaves throughout the canopy but prefer the upper foliage; larvae mine the leaves, leaving the epidermal areas on the mid/lower leaf surface unbroken [14]. Larvae move via cracks in the soil to find tubers, thus exposed tubers are pre-disposed to *P. operculella* damage [14, 50]. Larvae close to pupation drop to the ground and burrow into the tuber to complete its life cycle, making a swirl silk cocoon pupating on soil surface or in debris. Especially in warm dry climates, the larva can attack potato plants in field and storage causing great damage [96, 99].

3.1.4. Pupae

Occasionally, *P. operculella* pupae can be found on the surface of tubers (**Figure 6**), most commonly associated with tuber eyes [50]. *Phthorimaea operculella* pupae (0.84 cm long) are smooth and brown and often enclosed in a covering of fine residue that protects them from low temperatures and helps them endure the winter [76]. There is a clear distinction between male and female pupae. Rondon and Xue [81] evaluated the "scar" and the "width" method. Using the "scar" method, males could be recognized by the distance between the incision located between the 8th and 9th abdominal segment and the tip of the abdomen; there is also a gradual change in color eye pigmentation, which can help estimate the age of the pupae. Based on eye pigmentation, pupae are classified into newly formed pupa (yellowish in color, 1–2 day old pupae), followed by early red (3 day old), middle red (4 day old), late red, and black eye pupa (5–6 day old) [29, 81, 86, 102, 103]. Some studies suggest that the pupal period is not fixed but depends on the temperature at which the larvae grew [104]. Moregan and Crumb [101] reported 6–9 days as pupal period; Graft [20] reported 13–33 days; and Van der Goot [98] observed 14–17 days. Studies in the western USA indicated that *P. operculella* adults can potentially emerge from soil at depths up to 10 cm [76]. Once adults emerge, mating occurs, and within a few hours, females seek a potential host to lay their eggs.

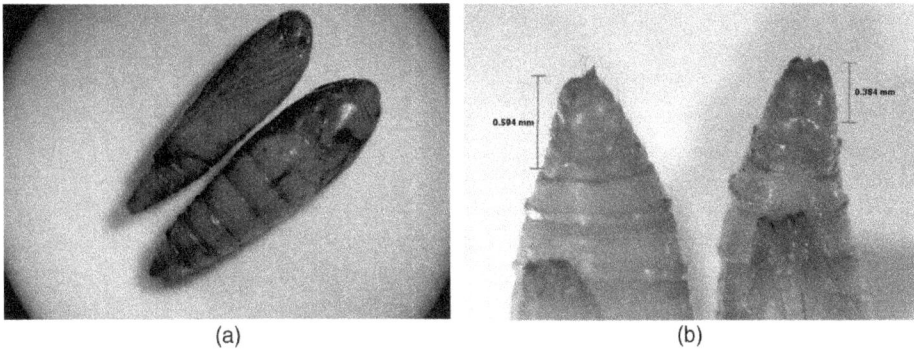

Figure 6. *Phthorimaea operculella* pupa and scar. (A) Female (right) and male (left); (B) Female (left); male (right).Photo credit: Oregon State University. Irrigated Agricultural Entomology Program (Rondon).

3.2. Life table

Phthorimaea operculella can complete several generations per year. Chittenden reported two generations of *P. operculella* in summer and a third generation in storage in the USA [105]; generally speaking, *P. operculella* is not a problem in the USA under controlled conditions [14]. In 2006, several potato storage controlled units were visited (n = 50), and only one had severe *P. operculella* infestation (Rondon personal observation). The infested unit stored tubers that came heavily infested from the field. Van der Goot [98] reported 6–8 generations a year in tropical regions; French [106] reported 2 generations in Australia, first in the winter and the second one on stored tubers; Graft [20], Trivedi and Rajagopal [65], and Sporleder et al. [107] reported 3–4 generations in Chile and the southern USA; Mukherjee [108] reported 13 generations per year in India, and Al-Ali et al. [24] reported 12 generations in Iraq. Recently, pheromone trapping in Bologna, Italy, where researchers integrated temperature dependent developmental time models, showed that *P. operculella* completed two generations throughout the potato-growing season; the remaining generations developed in the noncrop season [57]. This information suggests a correlation between geographical location, presence or absence of food source, and *P. operculella* generations per year [14]. Sporleder et al. [109] indicated that locations with one crop per season will have 2–3 generations per year (e.g., western USA), while locations with year-round crops like in India will have several generations per year [108].

3.3. Damage

Luscious, healthy, disease-free plants attract more *P. operculella* than wilting, nonirrigated plants [110]. Once *P. operculella* reaches a field, distribution of foliar damage tends to be nonrandom [7, 9, 111, 112] and more severe on the edges of the field facing the prevailing winds in a band parallel to the edge [9]. Larval density in foliage and tubers is higher at the margins of the field than in the center [18], a typical characteristic of pests that move from nearby areas [9, 17, 82]. Drier conditions in plants on field edges caused by wind and solar radiation leads to more *P. operculella* females looking for oviposition sites [17, 18, 70, 113]. Research shows that moths are able to forage beyond 100–250 m from center of origin [114]. In the western USA, most of the potatoes are vine-killed right before harvest; thus, when foliage is gone, *P. operculella* readily moves to nearby green fields or directly down to the tubers [50, 51].

3.4. Developmental thresholds

Phthorimaea operculella developmental threshold has been widely studied [29, 68, 88, 107, 109, 115, 116]. Developmental thresholds are necessary in order to establish best timing of control methods [97, 107, 117]. Differences in temperature for *P. operculella* development suggest the remarkable adaptation of this insect [14, 68, 69, 76, 107, 115]. Temperature-dependent development can be useful in forecasting occurrence and population dynamics of pests. Golizadeh and Zalucki [118] determined that the lower temperature threshold and thermal constant of immature stage were estimated to be 11.6°C and 338.5 degree days. A degree day is a measurement of heat units over time calculated from daily maximum and minimum temperatures; the minimum temperature at which insects' first start to develop is called the "lower developmental threshold," or baseline and the maximum temperature at which insects stop developing is called the "upper developmental threshold" or cutoff [119]. Golizadeh et al. [120] determined the average fecundity of females ranged from 45.3 eggs (at 16°C to 117.3 eggs (at 28°C); net reproductive rate (R_0) ranged from 12.8 (at 16°C) to 43.2 (at 28°C); and mean generation time (T) decreased with increasing temperatures from 61.0 days (at 16°C) to 16.2 days (at 32°C). These data suggest the close relationship between insect and abiotic factors.

3.5. Temperature affects life parameters of *P. operculella*

The developmental response of insects to temperature is important in understanding the ecology of insect life histories [121]. Temperature has an effect on geographical distributions, population dynamics, and management of insects [121]. For instance, studies by Langford and Cory [96] indicated that low temperatures retard and cause temporary cessation of *P. operculella* development not only physiologically but also due to the destructive effect of low temperatures on the food supply. Eggs exposed to 1.6–4.4°C for 4 months failed to hatch [96, 122]. Langford and Cory [96] indicated that outbreaks of *P. operculella* in Virginia in 1925 and 1930 coincided with hot and dry years, and the intensity of infestation varies in propor-tion to rainfall and humidity. Early studies in Maryland and Virginia [123, 124] indicated that *P. operculella* pupae can survive "short" constant sub-freezing temperatures. Several other authors reported that larvae and pupae could potentially survive frost [122, 125, 126]; other studies indicated that all life stages of *P. operculella* were killed by exposure to −6.6°C for 24 hours [96, 122]. Early studies by Langford in 1934 reported that *P. operculella* survived temperatures ranging from −11.6 to −6.6°C, but lengthy exposures to low temperatures were fatal to all stages. Trivedi and Rajagopal [65] found that pupae were extremely tolerant to low temperatures; however, Langford and Cory [96] indicated that full-grown larvae survived bet-ter at low temperatures [96]. In a manipulative study to determine how growth stage (egg, larva, or pupa) and soil depth affected the potential for winter survival, Döğramaci et al. [76] found that egg survival was reduced after 1 month of exposure to low temperatures; larvae were able to survive up to 30 days at 20-cm soil depth, while tubers at the surface buried at 6 cm were frozen; the pupal stage showed a greater tolerance to winter conditions (average −2°C) than the egg or larval stages, surviving up to 91 days of exposure. Hemmati et al. [78] studied the effect of cold acclimation. According to their study, super cooling points from 1st and 5th instar larvae, pre-pupae, and pupae were −21.8, −16.9, −18.9, and −18.0°C, respectively. Cold acclimation (1-week at 0 and 5°C) did not affect super cooling for 4–5th instar larvae, pre-pupae, and pupae. Also, LT_{50}s (lower lethal temperature for 50% mortality) for 1st and 5th instar larvae, pre-pupae, and pupae were −15.5, −12.4, −17.9, and −16.0°C, respectively. They concluded that cold acclimation resulted in a significant decrease in mortality of all developmental stages, and heat hardening also affects cold tolerance. A relatively recent study by Golizadeh et al. [120] determined that tubeworm failed to survive at 36°C during the egg period, and adult longevity was negatively correlated with temperature and the longest adult longevity was observed at 16°C.

3.6. Other parameters affecting *P. operculella*

Other parameters such as elevations and latitude seemed to play a role in *P. operculella* incidence [116]. Locations with higher spring, summer, or fall temperatures were associated with increased trapping rates in most seasons; in the western USA, trapping data from spring 2004 to fall 2005 showed that *P. operculella* males were present every week except in mid-January, with the greatest *P. operculella*/per trap occurring in December at around −0.09°C [50, 116, 127]; also, "warm" winters may also account for high *P. operculella* populations the following season [14, 116]. Similar observations were recorded in other insect species [128, 129]. In Israel, *P. operculella* first generation reached its peak in May or June (late spring, early summer) [18], and overlapping generations reached high numbers close to harvest which seems to be a characteristic of nondiapausing insects that continuously have access to host plants [18, 35, 104, 107, 110, 130, 131].

4. Monitoring

4.1. Pheromones

Phthorimaea operculella male moths are attracted to pheromones that are concentrated chemicals of the female "scent" impregnated in a rubber septum in the center of a sticky liner placed in delta traps [14]. According to Herman et al. [13], two chemicals have been identified as the main component of *P. operculella* sex pheromone: (E4, Z7)-tridecadienyl acetate (PTM1) [132] and (E4, Z7, Z10)-tridecatrienyl acetate (PTM2) [133]; chemicals have been synthesized, blended, and tested, and modifications are commercially used [134, 135]. Some other insects including other Gelechiidae moths could be trapped in the sticky liners; thus, liners should be changed once a week and lures should be changed once a month [50]. Pheromone traps are used to monitor populations in the field to help time insecticide applications [13]. Several authors found a positive relationship between the number of trapped adults and the density of larvae in the foliage and tuber [10, 75, 136]. Growers in areas impacted by *P. operculella* are encouraged to monitor insect using pheromone traps [50]; this has been an activity conducted in western USA states since 2005 (https://agpass.maps.arcgis.com/apps/webappviewer/index.html?id=8f3577c883ab4 ac58f262b4cd04ff569). Horne [137] used three methods to collected *P. operculella*: random and selected leaf samples and pheromone traps, concluding that random sampling of leaves did not always give adequate estimates as particular life stages could be overestimated or excluded from samples. In the western USA, pheromones traps are widely used [50]. Current recommendations include sampling at least 10 plants per field or section of the field; individual plants may be examined for the presence or absence of *P. operculella*; near 55% of the mines are found in the upper third of the potato plants are they are not easily to find; set at least 1 pheromone trap per 123 acres [50].

4.2. Action thresholds

Although treatment levels have not been established for *P. operculella*, California recommends a threshold of 15–20 moths per trap per night as a general threshold level [138] and 8 moths per trap per night for Oregon [139]. Keep in mind that *P. operculella* numbers vary from field to field and from area to area; thus, it is recommended to tailored management recommendations on field(s) specific information [50, 51]; and standard thresholds should be used exclusively as a reference.

4.3. Trapping

Kennedy [140], Bacon et al. [141], Raman [135], Salas et al. [142], and Tamhankar and Hawalkar [143] have reported results using different type of traps. In New Zealand, Herman et al. [13] tested water traps, which caught the greatest number of *P. operculella* per trap as compared to "DeSIRe" delta shaped sticky traps, "A traps" (cylinder-shaped), and funnel traps. Herman et al. [13] concluded that delta traps were the most suitable for commercial use. Coll et al. [18] presented information regarding pheromone traps plus poison bait placed on the ground at 50 m intervals in single rows with positive results. Based on Herman et al. [13] findings, the recommendation in the western USA has been to place at least one delta trap per potato field, beginning after canopy closure [50]; recent recommendations include placing four traps per field [14]. Soil type has an effect on number of moths caught per trap; thus, in sandy soils of Israel, pheromone traps caught almost twice as many moths than in loess fields [18].

5. Controlling *P. operculella*

Key aspects of the biology and ecology of *P. operculella* are important in selecting management practices to control this pest [14, 139]. Considering that most of the economic damage by this insect occurs when the insect infests the tubers, we should focus in early control of the pest [14]. For instance, deeper seed planting, hilling the rows, irrigation, and early harvest are a few of the methods suggested to prevent tuber infestation [10, 122, 126, 139, 144]. The use of chemicals, however, is still the main foundation of *P. operculella* control worldwide [139, 145–147]. It is advisable to check with your local extension or government agencies to review which pesticides are allowed to use in your region. Always read the labels and follow the manufacturers' recommendations.

5.1. Cultural control

Cultural methods reported to reduce *P. operculella* population include the elimination of cull piles and volunteers, timing of vine-kill, soil moisture at and after vine-kill, time between desiccation and harvest, rolling hills and covering hills, and cultivar selection [50, 139, 144]. In Tunisia, practices like deep seeding, hilling up, early harvest, irrigation until harvest, good sorting of tubers at harvest, and rapid harvesting prevent tuber infestation [148, 149]. In Sudan, planting date, planting depth, hilling-up, irrigation intervals, and mulching on insect infestation and on the greening of tubers in the field were studied [150]. Ali [150] indicated that tuber shape and skin characteristics had no effect on the degree of *P. operculella* infestation; early planting date resulted in fewer insect damage and greater yield compared to crops planted 3 weeks later; greater depth of planting and more frequent hilling-up significantly lowered infestation levels; light irrigation every 4 days and mulching with neem (*Azadirachta indica* L.) incorporated before harvest were the most effective treatments.

5.1.1. Elimination of volunteer potatoes and cull piles

The growth of volunteer potatoes is a serious problem because of the competition with current season crops but also for sanitary reasons [151] (**Figure 7**). For instance, Aarts and Sijtsma [152] indicated that volunteer potatoes can overgrow crops such as maize or sugar beets at planting, but lesser once established. In South Africa, *P. operculella* is described as a significant pest before harvest and during storage; and since eggs, larvae, or pupae can survive on volunteer potatoes, they represent a source of infestation for the following season [153, 154]. Besides volunteer potato elimination, cull piles should be removed to reduce overwintering stages which are a source of next years' population [75]. Certainly, in western USA states, volunteer potatoes can serve as a "green bridge" of numerous insect pests.

5.1.2. Rolling potatoes

Research has found that rolling of potato hills in sandy soil caused soil to slough off the hill, which resulted in increased *P. operculella* damage; obviously, this is not recommended in areas with sandy soils [50, 144]. Covering hills with 3–5 cm of soil immediately after vine-kill, which can be accomplished with a rotary corrugator, has been shown to significantly reduce tuber infestation [50, 144, 155]; however, it is not necessarily a common practice in the region. Others showed that exposed tubers are more prone to *P. operculella* infestation [9, 156]. They indicated that tuber infestation occurred 2–4 weeks before harvest and all infested tubers were covered with no more than 3 cm of soil.

5.1.3. Vine-killed

Tubers naturally mature as the potato plant senesces but tuber maturation can be artificially induced by killing the potato vines mechanically, chemically, or with a combination of both [14].

Figure 7. Volunteer potato in a crop field. Photo credit: Oregon State University. Irrigated Agricultural Entomology Program (Rondon).

Empirical observations suggest that all these activities have an impact on the level of *P. opercule-lla* infestation [14]. Field observations support the principle that *P. operculella* prefer green foliage to tubers for oviposition and feeding; thus, when foliage starts to decline, tubers are exposed, and therefore, infestations naturally increase; thus, the time between desiccation and harvest is crucial. The longer tubers are left in the field after desiccation, the greater the likelihood of tuber infestation [14, 50, 51]. Intuitively, tubers exposed or close to the soil surface are at high risk for *P. operculella* injury. In the Columbia Basin of Oregon, our recommendation includes to maintain more than 5 cm of soil over the tubers especially at the end of the season or after vine-killed [50].

5.1.4. Soil moisture

Phthorimaea operculela female moths favor dry soil for oviposition [70, 113]. Larval survivorship increased with decreasing soil moisture [113]. Then, keeping the soil moist to avoid cracks in the soil, particularly at the end of the season when vines are drying, reduces *P. operculella* tuber infestation. Rondon et al. [50], Clough et al. [155], and Rondon and Hèrve [139] researches have shown that irrigating daily with 0.25 cm through a center pivot irrigation system from vine kill until harvest decreased *P. operculella* tuber damage without increasing fungal or bacterial diseases. How water may decrease *P. operculella* movement? Since water closes soil cracks, reducing tuber access, *P. operculella* possibly perish from lack of oxygen in the soil due to water saturation, and/or their mobility is reduced by wet soil, decreasing their ability to move and find a tuber.

5.2. Biological control

Phthorimaea operculella is a relatively minor pest of potatoes in South America, probably due to the existence of a diverse complex of natural enemies attacking this pest [157]. Biological control, which is the use of living organisms to control pest populations, can have an environmental impact while controlling pest populations [158]. There are several organisms including parasitoids, and pathogens such as fungi or viruses, that have been used successfully to control *P. operculella* (**Table 3**).

5.2.1. Parasitoids

In general, Callan [159] indicated that the "newcomers" normally achieved better control than the "native" natural enemies, possibly caused by the long-term coevolution or adaptation between the relevant species at different trophic levels in South America. Thus, exploration for more parasitoids in the large parts of South America, Central America, and Mexico could be a promising way to improve the parasitoid-based biocontrol. Many parasitoids have been introduced mainly from South America, the area for the origin of *P. operculella* [153, 160]. Since 1918, biological control efforts attempted to introduce *Bracon gelichiae* L. to France from the Americas [36]. *Trichogramma* and *Copidosoma* species are among the most widely used parasitoids used to control *P. operculella* but with mixed results [69, 161–164]. *Copidosoma koehleri* Blanchard and *Bracon gelechiae* Ashmead have been used successfully in South America and Australia, respectively [66]; however, in Israel, *C. koehleri*, an encyrtid polyembryonic, did not reduce *P. operculella* populations, accounting only for 4–5% *P. operculella* larval reduction; similar results were found by Berlinger and Lebiush-Mordechi [165, 166] in Israel. In Italy, Pucci et al. [167] also found modest results. While several biotic and abiotic conditions in the eco-niche determine the population establishment and effectiveness of parasitoids [12], the reduction of human interference through reducing insecticide application is crucial for the parasitoid(s) establishment and performance [153]. Also, the effect of resources like flowering for parasitoids establishment is important [162]. In a laboratory study, when *C. koehleri* females were deprived of hosts for the first 5 days of their adult lives, neither the number of eggs laid nor longevity was significant affected [162]. Kfir [168] studied the fertility of *C. koehleri* compared to *Apanteles subandinus* L. under the effect of humidity in South Africa, concluding that low humidity is detrimental for the survivorship of this species. Choi et al. [169] and Aryal and Jung [170] reported *Diadegma fenestrale* L. found for the first time in Korea and accounted for 20–30% parasitism. In Sardinia, *Diadegma turcator* Aubert, and *Bracon nigricans* L., *B. properhebetor* L., and *Apanteles* spp. were also found attacking *P. operculella* [171].

Scientific name	Parasitizing stage	Place	Reference
Agathis gibbosa	Larvae	USA	Odebiyi and Oatman [172]
			Flanders and Oatman [173]
Agathis unicolor	Larvae	South America	Lloyd [157]
Apanteles litae	—	—	Lloyd [174]
Apanteles subandinus	Larvae	South Africa	Watmough et al. [153]
		South America	Franzmann [175]
		Australia	Horne [176]
		India	Rao and Nagaraja [177]
		Zimbabwe	Mitchell [178]
Apanteles scutellaris	Larvae	India	Rao and Nagaraja [177]
		USA	Flanders and Oatman [173]
Bracon gelechiae	—	India	Rao and Nagaraja [177]
Bracon hebator		India	Divakar and Pawar [179]
Campoplex haywardi	Larvae	USA	Leong and Oatman [180]
Campoplex phthorimaeae	—	—	Flanders and Oatman [173]
Chelonus blackburni	Eggs, larvae	India	Choudhary et al. [181]
Chelonus contractus	Eggs	France	Labeyrie [182]
Chelonus curvimaculatus	Eggs	South Africa	Watmough et al. [153]
Chelonus kellieae	Larvae	USA	Flanders and Oatman [173]
			Powers [183]
Chelonus phthorimaea	Larvae	USA	Powers [183]
Copidosoma desantis	Eggs	Australia	Franzmann [175]
Copidosoma koehleri	Larvae	Australia	Horne [176]
		South Africa	Keasar [184]
		Israel	Cruickshank and Ahmed [185]
		South America	Kfir [186]
		Italy	Lloyd [157]
		Zimbabwe	Mitchell [178]
Copidosoma uruguayensis	Larvae	South Africa	Watmough et al. [153]
		South America	
Diadegma compressum	—	—	Flanders and Oatman [173]
Diadegma molliplum	—	—	Lloyd and Guido [174]
Diadegma stellenboschense	Larvae	South Africa	Watmough et al. [153]
Elasmus funereus	Larvae	Australia	Franzmann [175]
Habrobracon gelechiae	Larvae	US Pacific Northwest	Rondon 2007 [50]
Microchelonus curvimaculatus	Larvae	Australia	Franzmann [175]
Microgaster phthorimaea	—	—	Flanders and Oatman [173]

Scientific name	Parasitizing stage	Place	Reference
Nepiera fuscifemora	—	—	Flanders and Oatman [173]
Orgilus californicus	—	—	Flanders and Oatman [173]
Orgilus lepidus	Larvae	Australia	Franzmann [175]
			Horne [176]
Orgilus parcus	Larvae	South Africa	Watmough et al. [153]
Orgilus jenniae	Larvae	USA	Flanders and Oatman [187]
Parahormius pallidipes	—	—	Flanders and Oatman [173]
Pristomerus spinator	—	—	Flanders and Oatman [173]
Sympiesis stigmatipennis	—	—	Flanders and Oatman [173]
Temelucha minuta	Larvae	Australia	Franzmann [175]
Temelucha picta	Larvae	South Africa	Watmough et al. [153]
Trichogramma brasiliensis	Eggs	India	Harwalkar and Rananavere [188]
Zagrammosoma flavolineatum	—	—	Flanders and Oatman [173]

Table 3. List of parasitoids that control *Phthorimaea operculella* Zeller. Compiled by Y. Gao.

5.2.2. Predators

The role of generalist predators such as *Orius* spp. [189], hymenopteran [190], *Dicranolaius* spp. [191], phytoseiid [192], *Chrysoperla* spp. [193], *Agistemus* [194], and others predators present in potato ecosystems has not been widely studied [18]. *Geocoris* sp. (Hemiptera: Miridae) was seen as a potential *P. operculella* predator in Nepal [195].

5.2.3. Biorationals

Other potential biological control agents include nematodes and entomopathogens. The nematode of the genus *Hexamermis, Steinernema,* and *Heterorhabditis* are suggested to exert significant control on *P. operculella* [196, 197]. *Steinernema feltiae, S. bibionis, S. carpocapsae,* and *Heterorhabditis heliothidis* were used in laboratory experiments in Russia with promising results [198]. Kakhki et al. [199] found that the higher the concentration (0, 75, 150, 250, 375, and 500 js/mL) of *S. carpocapsae* and *H. bacteriophora,* the higher is the mortality in both larval and pre-pupal stages. Fungal entomopathogens such as *Beauveria bassiana* and *Metarhizium anisopliae* have been isolated from *P. operculella* larvae, extracted, and used as bio-insecticides causing *P. operculella* death at a rate higher than 80% [200–204]. Back in 1967, a granulovirus was found and reported as a new record [205]; the following years, the granulovirus was isolated and collectively named *Phthorimaea operculella granulovirus* (PhopGNV). They are well known for efficiently controlling and preventing *P. operculella* in storage [206]. Since first reported, GNV has been tested for pest control in the fields in South America and Australia [206]. Arthurs et al. [207] evaluated PoGV and *Bacillus thuringiensis subsp. kurstaki* for control of *P. operculella* in stored tubers with limited efficacy.

5.3. Chemical control

5.3.1. Field

Traditional chemical control targeting mainly larvae and adults is well documented [208, 209]. Back in the 1970s, azinphos-ethyl and endosulfan were effective against foliage mining [7]. Others reported thiacloprid, quinalphos, and diflubenzuron as effective [210, 211]. Rondon et al. [50] provided some information related to pesticide use in the USA. However, potential strategies to improve chemical control are also being investigated. Mahdavi et al. [212] studied the insecticidal activity of plant essential oils including the insecticidal and residual effects of nanofiber oil and pure essential oil of *Cinnamomum zeylanicum* L. under laboratory conditions; fumigant toxicity was evaluated on different growth stages (egg, male, and female adults) of *P. operculella* with encouraging results. Similarly, Mahdavi et al. [213] tested *Zingiber officinale* Roscoe, demonstrating the relative effectiveness of additional means of control. Tanasković et al. [214] revised the effect of several plants as bio-insecticides to suppress *P. operculella*.

5.3.2. Storage

In the USA, *P. operculella* is not a problem during storage, since storage occurs under controlled conditions of temperature and humidity [14]. However, in other parts of the world, *P. operculella* can cause significant damage during storage. Moawad and Ami [215] and Abewoy [30] reported that *P. operculella* causes serious damage to stored potato through its larval tunneling and feeding, which can lead to secondary infection by fungi or bacteria. During storage, the damaged tubers become unsuitable for human consumption; moreover, the adult moth flies from the infested tubers in the storage and from neglected warehouses or farms back to the fields, where it causes pre-harvest infestation. Granulovirus was found to efficiently control *P. operculella* in Colombia and was used as a biopesticide in storage conditions [206]. Early on, Raman and Booth [216], Raman et al. [217], and Lal [218, 219] indicated that *P. operculella* could be reduced by covering tubers with *Lantana camara* L., *L. aculeate* L., or *Eucalyptus globulus* Labill foliage. Niroula and Vaidya [220] reported good control using *Minthostachys* spp., *Baccharis* spp., *L. neesiana*, and *Artemisia calamus* L.; Sharaby et al. [221] tested peppermint oils, camphor, eugenol, and camphene in Egypt.

5.4. Resistance: plant versus insect resistance?

Host plant resistance enables plants to avoid, tolerate, or recover from pest infestations [222, 223]. Genome diversity of tuber-bearing potato presents a complex evolutionary history that complicates domestication in the cultivated potato [224]. Currently, abundant evidences in other insect-plant interactions systems exist, especially the defensive chemical compounds. In contrast, the amount of information to improve potato genotypes against *P. operculella* is still lacking [225]. Cultivated potatoes have more than 100 tuber-bearing relatives native to the Andes of southern Peru; among them, *Solanum chiquidenum* L. and *Solanum sandemanii* L. for instance, which are highly resistant to *P. operculella*, damage in tubers [226]. The nutritional value of the host is an important resistance factor limiting normal growth and development of *P. operculella* [227]. Moreover, some potato hybrids can inhibit oviposition, while surviving larvae had higher mortality and slower feeding rates than those larvae reared on foliage of cultivated potatoes [228]. An improved investigation of the mechanisms for the traits associated with the tuber and foliage resistance and the introduction of these traits into commercial varieties may be an effective way to enhance the plant resistance against *P. operculella*. Golizadeh et al. [229] tested the resistance of six potato cultivars; also, Rondon et al. [85] studied potato lines, some of which exhibit promising results for controlling mines and number of larvae in potato tubers [77]. An earlier study by Rondon et al. [85] confirmed that tubers of the transgenic clone Spunta G2 were resistant to *P. operculella* damage. Spunta G2 was developed in the early 2000s [230, 231]. In recent years, plants have received genes that encode toxic proteins to resist against insects [232, 233]. Thus, researchers like Fatehi et al. [234] evaluated the effect of wheat extracts against digestive alpha-amylase and protease activities against *P. operculella*; those enzymes are important digestive enzymes used during the feeding process. Inhibition of enzymes could potentially help us reduce or stop *P. operculella* feeding. Also, radiation to induce sterility of *P. operculella* males has also been studied [235–237].

6. Conclusions

Phthorimaea operculella is considered one of the most important potato pests worldwide. It is a cosmopolitan pest of solanaceous crops including potato, tomato, and other important row crops. Based on *P. operculella* biology, ecology, including its relationship with the potato crop well thoughout pest management practices, can keep the pest under control. The effectiveness of control methods will depend on the response time to pest infestation, resources available, and also, pest management practitioner experience. This chapter includes up-to-date information related to *P. operculella* that we anticipate will be useful to growers, fieldmen, and producers that face the challenges imposed by this pest.

Acknowledgements

The authors thank Dr. Lukas, Oregon State University for reviewing an early version of the manuscript. In addition, authors thank to Dr. James Crosslin, retired USDA ARS scientist for his editing; also thanks to Ira D. Thompson and Maria Montes de Oca for proofing.

Author details

Silvia I. Rondon[1]* and Yulin Gao[2]

*Address all correspondence to: silvia.rondon@oregonstate.edu

1 Hermiston Agricultural Research and Extension Center, Oregon State University, Hermiston, Oregan, USA

2 Institute of Plant Protection, Chinese Academy of Agricultural Sciences, Beijing, P.R. China

References

[1] Andre CM, Ghislain M, Bertin P, Oufir M, del Rosario Herrera M. Andean potato cultivars as a source of antioxidant and mineral nutrient. Journal of Agriculture and Food Chemistry. 2007;**55**:366-378

[2] Navarre DA, Goyer A, Shakya R. Nutritional value of potatoes: Vitamin, phytonutrient and mineral content. In: Singh J, Kaur L, editors. Advances in Potato Chemistry and Technology. Elsevier Inc; 2009

[3] Goyer A, Navarre DA. Folate is higher in developmentally younger potato tubers. Journal of the Science of Food and Agriculture. 2009;**89**:579-583

[4] Robinson BR, Sathuvali V, Bamberg J, Goyer A. Exploring folate diversity in wild and primitive potatoes for modern crop improvement. Genes. 2015;**6**(4):1300-1314

[5] Johnson DA. Potato Health Management. 2nd ed. USA: APS St Paul Minnesota; 2008. pp. 258

[6] Bacon OG. Systemic insecticides applied to cut seed pieces and to soil at planting time to control potato insects. Journal of Economic Entomology. 1960;**53**:835-839

[7] Foot MA. Field assessment of several insecticides against the potato tuber moth *Phthorimaea operculella* (Zell.) at Pukukohe. New Zealand Journal of Experimental Agriculture. 1974;**2**:191-197

[8] Haines CP. The potato tuber moth, *Phthorimaea operculella* (Zeller): A bibliography of recent literature and a review of its biology and control on potatoes in the field and in store. Tropical Products Institute. 1977;**G112**(3). 15 pp

[9] Foot MA. Bionomics of the potato tuber moth, *Phthorimaea operculella* (Lepidoptera: Gelechiidae) at Pukekohe. New Zealand Journal of Zoology. 1979;**6**:623-636

[10] Shelton AM, Wyman JA. Time of tuber infestation and relationships between catches of adult moths, foliar larval populations, and tuber damage by potato tuber worm. Journal of Economic Entomology. 1979;**72**:599-601

[11] Shelton AM, Wyman JA. Potato tuberworm damage to potato grown under different irrigation and cultural practices. Journal of Economic Entomology. 1979;**72**:261-264

[12] Briese DT. Geographic variability in demographic performance of the potato moth, *Phthorimaea operculella* Zell. Australian Bulletin of Entomological Research. 1986;**76**:719-726

[13] Herman TJB, Clearwater JR, Triggs CM. Impact of pheromone trap design, placement and pheromone blend on catch of potato tuber moth. New Zealand Plant Protection. 2005;**58**:219-223

[14] Rondon SI. The potato tuberworm: A literature review of its biology, ecology, and control. American Journal of Potato Research. 2010;**87**:149-166

[15] Fenemore PG. Host-plant location and selection by adult potato moth, *Phthorimaea operculella* (Lepidoptera: Gelechiidae): A review. Journal of Insect Physiology. 1988;**34**:175-177

[16] Das GP, Raman KV. Alternate hosts of the potato tuber moth, *Phthorimaea operculella* (Zeller). Crop Protection. 1994;**13**:83-86

[17] Gilboa S, Podoler H. Presence-absence sequential sampling for potato tuberworm (Lepidoptera: Gelechiidae) on processing tomatoes: Selection of sample sites according to predictable seasonal trends. Journal of Economic Entomology. 1995;**88**:1332-1336

[18] Coll M, Gavish S, Dori I. Population biology of the potato tubermoth, *Phthorimaea opercuella* (Lepidoptera: Gelechiidae) in two potato cropping systems in. Israeli Bulleton Entomological Research. 2000;**90**:309-315

[19] Keller S. Integrated pest management of the potato tuber moth in cropping systems of different agro-ecological zones. In: Kroschel J, editor. Advances in Crop Research. Margraft Verlag; 2003. 153 pp

[20] Graft JE. The potato tubermoth. Technical Bulletin USDA. 1917;**427**: 58 pp

[21] Balachowsky AS, Real P. La teigne de la pomme de terre. In: Balachoswky AS, editor. Entomologie appliquee a l'agriculture, Tome II, Lepidopteres. 1966;**i**:371-381

[22] Arnone S, Musmeci S, Bacchetta L, Cordischi L, Pucci E, Cristofaro M, et al. Research in *Solanum* spp. as sources of resistance to the potato tuber moth *Phthorimaea operculella* (Zeller). Potato Research. 1998;**41**:39-49

[23] Fadli HA, Al-Salih GAW, Abdul-Masih AE. A survey of the potato tuber moth in Iraq. Journal Iraqi Agriculture. 1974;**29**:35-37

[24] Al-Ali AS, Al-Neamy IK, Abbas SA. Observations on the biology of the potato tuber moth *Phthorimaea operculella* (Zeller) (Lepidoptera: Gelechiidae) in Iraq. Zeitschrift für Angewandte Entomologie. 1975;**79**:345-351

[25] Lall BS. Preliminary observations on the bionomics of potato tuber-moth (*Gnorimoschema opercullela* Zell.) and its control in Bihar, India. Indian Journal of agricultural Science. 1949;**19**:295-306

[26] Nirula KK. Control of potato tuber-moth. Indian Potato Journal. 1960;**2**:47-51

[27] Nirula KK, Kumar R. Control of potato tuber-moth in country stores. Indian Potato Journal. 1964;**6**:30-33

[28] Gubbaiah A, Thontadarya TS. Bionomics of potato tuberworm, *Gnorimoschema operculella* Zeller (Lepidoptera Gelechiidae) in Karnataka. Mysore Journal of Agricultural Science. 1977;**11**:380-386

[29] Chauhan U, Verma LR. Biology of potato tuber moth *Phthorimaea operculella* Zeller with special reference to pupal eye pigmentation and adult sexual dimorphism. Journal of Economic Entomology. 1991;**16**:63-67

[30] Abewoy D. Review on potato late blight and potato tuber moth and their integrated pest management options in Ethiopia. Advances in Crop Science Technology. 2017;6(1):331

[31] Lagnaoui A, Cañedo V, Douches D. Evaluation of Bt-cry1Ia1 (cry V Transgenic potatoes) on two species of potato tuber moth, Phthorimaea operculella and Symmestrischemas tangolis (Lepidoptera: Gelechiidae). CIP program report, Peru; 2000. pp 117-121

[32] Zeller PC. Beitrage Zur Kenntniss der nordamericanishchen Nachtfolter, besonders der Microlepidopteran. Verhandlungen der Zoologish-botanishchen Geselischaft in Wein. 1873;23:262-263

[33] Meyrick E. A new genus of Gelechiidae. Entomological Magazine. 1902;38:103-104

[34] Povolny D, Weismann L. *Kritischer Beitrag zur Problematik der Ruben Phthorimaea operculella (Zeller)*. Folia Zoologica. 1958;8:97-121

[35] Povolny D. Gnorimoschemini trib. nov. eine neie Tribus der familie Gelechiidae nebst Bemerkungen zu ibrer taxonomic (Lepidoptera). Cassopis Ceskoslovenske Spolecnosti Entomologicke. 1964;61:330-359

[36] Center for Agriculture and Biosciences International, CABI, https://www.cabi.org/. CABI (Centre for Agriculture and Biosciences International). The Crop Protection Compendium. 2010th ed. Wallingford, UK: CABI Publishing; 2010. www.cabi.org/cpc2010

[37] Entomological Society of America Database. Available from: https://www.entsoc.org/sites/default/files/files/common_name.pdf

[38] Barragan AR. Identificacion, biologia, y comportamiento de las polillas de la papa en el Ecuador. PROMSA-MAG, PUCE. 2005;1:1-11

[39] European and Mediterranean Plant Protection Organization, 38 EPPO Global Database. Available from: https://gd.eppo.int/ taxon/PHTOOP

[40] Flint M. Integrated Pest Management For Potatoes in the Western United States. Pub. 3316. Univ. of Cal 1986. pp. 1-146

[41] Rothschild GHL. The potato moth: An adaptable pest of short term cropping systems. In: Kitching RL, editor. The Ecology of Exotic Plants and Animals. Brisbane: J. Wiley; 1986. pp. 144-162

[42] Kroschel J, Koch W. Studies on the population dynamics of the potato tuber moth *Phthorimaea operculella* Zeller (Lepidoptera: Gelechiidae) in the Republic of Yemen. Journal of Applied Entomology. 1994;118:327-341

[43] Fuglie K, Salah B, Essamet M, Ben Temine A, Rahmouni A. The development and adoption of integrated pest management of the potato tuber moth *Phthorimaea operculella* (Zeller) in Tunisia. International Journal of Tropical Insect Science. 1992;14:501-509

[44] Visser D. The Potato Tuber Moth *Phthorimaea operculella* (Zeller) in South Africa: Potential Control Measures in Non-rRefrigerated Store Environments. University of Pretoria; 2004. pp. 60

[45] Berthon CH. On the potato moth. Proceedings of the Royal Society of Van Diemen's Land. 1855;**3**:76-80

[46] Lefroy HM. The potato tuber moth. Indian Agricultural Journal. 1907;**2**:294-295

[47] Chittenden FH. The potato tubermoth. United States Department of Agriculture Farmer's Bulletin. 1913;1-7

[48] Radcliffe EB. Insect pest of potato. Annual Review of Entomology. 1982;**27**:173-204

[49] Jensen A, Hamm PB, Schreiber A, DeBano S. Prepare for tuber moth in 2005. Potato Progress. 2005;**5**:1-4

[50] Rondon SI, DeBano SJ, Clough GH, Hamm PB, Jensen A, Schreiber A, et al. Biology and management of the potato tuberworm in the Pacific Northwest. PNW 594; 2007

[51] Rondon SI, Schreiber A, Hamm PB, Wohleb C, Waters T, Cooper R, et al. Potato psyllid vector of zebra chip disease in the PNW. 633. PNW Extension publication;2017

[52] Chen JB. Observation of the tobacco moth and its prevention in Guangxi. Interesting insects. 1937;**2**:15

[53] Xu SY. Control Methods of Tobacco Pests. Zhenzhou, China: Henan Science and Technology Press; 1985

[54] Hu J. Occurrence and control method of the potato tuberworm. Plant Doctor. 2008;**21**:46

[55] Whilshire EP. The Lepidoptera of Iraq. Ministry of Agricolture, Government of Iraq; 1957. 160 pp

[56] Zagulyaev A. Potato tuber moth, *Phthorimaea operculella* Zeller (Lepidoptera: Gelechiidae). Entomologiceskoe Obozrenie. 1982;**61**:817-820

[57] Masetti A, Butturini A, Lanzoni A, DeLuigi V, Burgio G. Area-wide monitoring of potato tuberworm (Phthorimae operculella) by pheromone trapping in Northern Italy: Phenology, spatial distribution and relationships between catches and tuber damage. Agricultural and Forest Entomology. 2014;**17**(2):138-145

[58] Medina RF, Rondon SI. Population structure of the potato tuberworm Phthorimaea operculella Zeller (Lepidoptera: Gelechiidae) in the United States. In: 6th Annual Meeting of the Entomological Society of America. Nov. 16-19. Reno, NV; 2014

[59] Clark JF. New species of micrlepidoptera from Japan. Entomology News. 1962;**73**:102

[60] Urbaneja A, Vercher R, Navarro V, Garcia MF, Porcuna JL. La polilla del tomate, *Tuta absoluta*. Phytoma Esspana. 2007;**194**:16-23

[61] Fenemore PG. Oviposition of potato tuber moth, *Phthorimaea operculella* Zell. (Lepidoptera: Gelechiidae); identification of host-plant factors influencing oviposition response. New Zealand Journal of Zoology. 1980;**7**(3):435-439

[62] Shands WA, Allen N, Gilmore JW. A survey of insect injury to tobacco grow for the flue curing. Journal of Economic Entomology. 1938;**3**:116-117

[63] Bartoloni P. La *Phthorimaea operculella* Zeller (Lep. Gelechiidae) in Italia. Redia. 1951; **36**:301-379

[64] Cunningham IC. Alternative host plants of tobacco leaf miner (*Phthorimaea operculella* Zeller). Queensland Journal Agriculture and Animal Science. 1969;**26**:107-111

[65] Trivedi TP, Rajagopal D. Distribution, biology, ecology and management of potato tuber moth, *Phthorimaea operculella* (Zeller) (Lepidoptera: Gelechiidae): A review. Tropical Pest Management. 1992;**38**:279-285

[66] Alvarez JM, Dotseth E, Nolte P. Potato Tuber Worm a Threat for Idaho Potatoes. Moscow, ID: University of Idaho Extension, Idaho Agricultural Experiment Station; 2005. (updated 2014)

[67] Morris HM. Potato tuber moth (*Phthorimaea operculella* Zell.). Cyprus Agricultural Journal. 1933;**28**:111-115

[68] Attia R, Mattar B. Some notes on the potato tubermoth *Phthorimaea operculella* Zell. Bulletin of the Society of Entomology Egypt. 1939;**216**:136

[69] Broodryk SW. The biology of *Diadegma stellenboschense* (Cameron) [Hymenoptera: Ichneumonidae], a parasitoid of potato tuber moth. Journal of the Entomological Society of Southern Africa. 1971;**34**:413-423

[70] Meisner J, Ascher KRS, Lavie D. Factors influencing the attraction to the potato tuber moth, *Gnorimoschema operculella* Zeller. Journal of Applied Entomology. 1974;**77**:179-189

[71] Kroschel J. Integrated Pest Management in Potato Production in Yemen with Special Reference to the Integrated Biological Control of the Potato Tuber Moth (Phthorimaea operculella Zeller). Tropical Agriculture, 8. Weikersheim, Germany: Margraf Verlag; 1995

[72] Varela LG, Bernays EA. Behavior of newly hatched potato tuber moth larvae, *Phthorimaea operculella* Zell. (Lepidoptera: Gelechiidae), in relation to their host plants. Journal of Insect Behavior. 1988;**1**:261-275

[73] Futuyma DJ. Potential evolution of host range in herbivorous insects. In: Proceedings of the Host specificity testing of exotic arthropods biological control agents: The biological basis for improvement in safety; 2000. pp. 42-53

[74] Hitchner EM, Kuhar TP, Dickens JC, Youngman RR, Schultz PB, Pfeiffer DG. Host plant choice experiments of Coloroado potato beetle (Coleoptera: Chrysomelidae) in Virginia. Journal of Economic Entomology. 2008;**101**(3):859-865

[75] Shelton AM, Wyman JA. Post-harvest potato tuberworm Lepidoptera, Gelechiidae population-levels in cull and volunteer potatoes, and means for control. Journal of Economic Entomology. 1980;**73**:8-11

[76] Dögramaci M, Rondon SI, DeBano SJ. The effect of soil depth and exposure to winter conditions on survival of the potato tuberworm *Phthorimaea operculella* (Lepidoptera: Gelechiidae). Entomologia Experimentalis et Applicata. 2008;**129**:332-339

[77] Gill HK, Chahil G, Goyal G, Gill AK, Gillett-Kaufman JL. Potato tuberworm Phhtorimaea operculella (Zeller) (Lepitopera: Gelechiidae). EDIS IFAS Extension EENY 587. Rev 2017; 2014

[78] Hemmati C, Moharramipour S, Asghar Talebi A. Effects of cold acclimation, cooling rate, and heat stree on cold tolerace of the potato tuber moth *Phthorimaea operculella* (Lepidoptera: Gelechiidae). European Journal of Entomology. 2014;**111**(4):487-494

[79] Van Lenteren JC, Nodus LPJJ. Whitefly plant relationships, behavioral and ecological aspects. In: Gerling D, editor. White Flies: Their Bionomics, Pest Status, and Management. Intercept Limited. 1990. pp. 47-89

[80] Raman KV. The Potato Tuber Moth. Technical Information Bulletin 3. Peru: International Potato Center Lima; 1980 (Revised edition 1980)

[81] Rondon SI, Xue L. Practical techniques and accuracy for sexing the potato tuberworm, *Phthorimaea operculella* (Lepidoptera: Gelechiidae). Florida Entomology. 2010; **93**(1):113-115

[82] Reed EM. Factors affecting the status of a virus as a control agent for the potato moth (*Phthorimaea operculella* Zell.) (Lepidoptera: Gelechiidae). Bulletin of Entomological Research. 1971;**61**:207-222

[83] Foley DH. Tethered flight of the potato moth, *Phthorimaea operculella*. Physiological Entomology. 1985;**10**:45-51

[84] Krambias A. Climactic factors affecting the catches of potato tuberworm, *Phthorimaea operculella* (Zeller) at a pheromone trap. Bulletin of Entomological Research. 1976;**66**:81-85

[85] Rondon SI, Hane D, Brown CR, Vales MI, Dŏgramaci M. Resistance of potato germplasm to the potato tuberworm (Lepidoptera: Gelechiidae). Journal of Econonomic Entomology. 2009;**102**(4):1649-1653

[86] Chauhan U, Verma LR. Adult eclosion and mating behaviour of potato tubermoth, *Phthorimaea operculella* Zeller. Journal of the Indian Potato Association. 1985;**12**:148-157

[87] Makee H, Saour G. Factors influencing mating success, mating frequency, and fecundity in *Phthorimaea operculella* (Lepidoptera: Gelechiidae). Environmental Entomology. 2001;**30**:31-36

[88] Broodryk SW. Dimensions and developmental values for potato tuber moth *Phthorimaea operculella* (Zeller) in South Africa. Phytophylactica. 1970;**2**:215-216

[89] Traynier RM. Field and laboratory experiments on the site of oviposition by the potato moth. Bulletin of Entomological Research. 1975;**65**:391-398

[90] Traynier RM. Influence of plant and adult food and fecundity of potato tuber moth, *Phthorimaea operculella*. Experimental and Applied Entomology. 1983;**33**:145-154

[91] Cannon RC. Investigations in the control of the potato tuber moth, *Gnorimoschema operculella* Zell. (Lepidoptera: Gelechiidae) in north Queensland. Queesnland Journal of Agricultural Science. 1948;**5**:107-124

[92] Trehan KN, Bagal SR. Life history and bionomics of potato tuber moth *Phthorimaea operculella* Zell. (Lepidoptera: Gelechiidae). Procedings of the Indian Acadamy of Science. 1944;**19**:176-187

[93] Labeyrie V. Influence de l'alimentation sur la ponte de la teigne de la pomme de terre (*Gnorimoschema operculella* Z.) (Lep. Gelechiidae). Bulletin of the Society of Entomolgy of France. 1957;**62**:64-67

[94] Makee H, Saour G. Nonrecovery of fertility in partially sterile male *Phthorimaea operculella* (Lepidoptera: Gelechiidae). Journal of Economic Entomolgy. 1999;**92**:516-520

[95] Van Loon JJA. Insect-Host Integrations: Signals, Sense, and Selection Behavior. Wageningen University; 2013. pp. 1-28. Avaiable from: http://edepot.wur.nl/330155

[96] Langford GS, Cory EN. Winter survival of the potato tuber moth, *Phthorimaea operculella* (Zeller). Journal of Economic Entomology. 1934;**27**:210-213

[97] Ascerno M. Insect phenology and integrated pest management. Journal of Arboriculture. 1991;**17**:13-15

[98] Van der Goot P. Brestrisding Van de aardappel-Knolrups in Goedangs. Korte Meded. Inst. Piziektenziekten. 1926;**1**:17

[99] Stanev M, Kaitazov A. Studies on the bionomics and ecology of the potato moth *Gnorimoschema (Phthorimaea) operculella* Zeller in Bulgaria and means for its control. Izv. nauch. Inst. Zasht. Rast. 1962;**3**:49-89

[100] Verma RS. Bionomics of *Gnorimoschema operculella* Zeller (Lepidoptera: Gelechiidae). Labdev: Journal of Science and Technology. 1967;**5**:318-324

[101] Moregan AC, Crumb SE. The tobacco split worm. Bulletin of the US Department of Agriculture. 1914;**59**:7

[102] Larraín PS, Guillon M, Kalazich J, Graña F, Vásquez C. Effect of pheromone trap density on mass trapping of make potato tuber moth *Phthorimaea operculella* (Zeller) (Lepidoptera: Gelechiidae) and level of damage on potato tubers. Chilean Journal of Agricultural Research. 2009;**69**:281-285

[103] Summers KM, Howells AJ, Pyliotis NA. Biology of eye pigmentation in insects. Advances in Insect Physiology. 1982;**16**:119-166

[104] Whiteside EF. An adaptation to overwintering in the potato tuber moth, *Phthorimaea operculella* (Zeller) (Lepidoptera: Gelechiidae). Journal of the Entomological Society of Southern Africa. 1985;**48**:163-167

[105] Chittenden FH. The potato tuber moth (*Phthorimaea operculella* Zeller). United State Department of Agriculture. 1912;**162**:5

[106] French JG. The potato moth, *Phthorimaea operculella* Zeller. Recent spraying experiments in Gippsland. Journal of the Deptartment of Agriculture Victoria. 1915;**23**:6144-6180

[107] Sporleder M, Kroschel J, Quispe MRG, Lagnaoui A. A temperature-based simulation model for the potato tuberworm, *Phthorimaea operculella* Zeller (Lepidoptera: Gelechiidae). Environmental Entomology. 2004;**33**:477-486

[108] Mukherjee AK. Life-history and bionomics of potato moth (*Gnorimoschema operculella* Zell.) at Allahabad (U.P.) together with some notes on the external morphology of the immature stages. Journal of the Zoological Society of India. 1949;**1**:57-67

[109] Sporleder M, Simon R, Juarez H, Kroschel J. Regional and seasonal forecasting of the potato tuber moth using a temperature-driven phenology model linked with geographic information systems. In: Kroschel J, Lacey L, editors. Integrated Pest Management for the Potato Tuber Moth, *Phthorimaea operculella* Zeller - a Potato Pest of Global Importance. Tropical Agriculture 20, Advances in Crop Research 10. Weikersheim, Germany: Margraf Publishers; 2008

[110] Yathom S. Phenology of the tuber moth, *Gnorimoschema operculella* Zell., in Israel in the spring. Israel Journal of Agricltural Research. 1968;**18**:89-90

[111] Foot MA. Cultural practices in relation to infestation of potato crops by the potato tuber moth (*Phthorimaea operculella*). II. Effect of seed depth, re-moulding, pre-harvest defoliation, and delayed harvest. New Zealand Journal of Experimental Agriculture. 1976;**4**:121-124

[112] Foot MA. Cultural practices in relation to infestation of potato crops by the potato tuber moth (*Phthorimaea operculella*). I. Effect of irrigation and ridge width. New Zealand Journal of Experimental Agriculture. 1974;**2**:447-450

[113] Meisner J. Atraction and repelllence in the potato tuber moth, Gnorismoschema operculella Zeller: Phagostimulants and antifeedants for the larvae: Some of the factors to oviposition [PhD thesis]. Jerusalem, Israel: The Hebrew University; 1969

[114] Cameron PJ, Walker GP, Penny GM, Wingley PJ. Movement of potato tuberworm (Lepidoptera: Gelechiidae) within and between crops, and some comparisions with diamonback moth (Lepidoptera: Pluteliidae). Environmental Entomology. 2002;**31**:65-75

[115] Davoud MA, El-Saadany GB, Mariy FMA, Ibrahim MY. The thermal threshold units for *Phthorimaea operculella* (Zeller). Annals of Agricultural Science. 1999;**44**:379-393

[116] DeBano SJ, Hamm PB, Jensen A, Rondon SI, Landolt PJ. Spatial and temporal dynamics of the potato tuberworm (Lepidoptera: Gelechiidae) in the Columbia Basin of the Pacific Northwest. Journal of Economic Entomology. 2010;**39**(1):1-14

[117] Legg DE, Van Vleet SM, Lloyd JE. Simulated predictions of insect phenological events made by using mean and median functional lower developmental thresholds. Journal of Economic Entomology. 2000;**93**:658-661

[118] Golizadeh A, Zalucki MP. Estimating temperature dependent developmental rates of potato tuberworm, *Phthorimaea operculella* (Lepidoptera: Gelechiidae). Insect Science. 2012;**19**:609-620

[119] Herms DA. Using degree-days and plant phenology to predict pest activity. In: IPM of midewest lasncapes. MN Agriculture Experimental Statio; 2004. pp. 49-59

[120] Golizadeh A, Razmjou J, Rafuee-Dastjerdi H, Hassanpour M. Effects of temperature on development, survival, and fecundity of potato tuberworm *Phthorimaea operculella* on potato tubers. American Journal of Potato Research. 2012;**89**(2):150-158

[121] Régnière J, Powell J, Bentz B, Nealis V. Effects of temperature on development, survival, and reproduction of insects: Experimental design, data analysis, and modeling. Journal of Insect Physiology. 2012;**58**(5):634-647

[122] Langford GS, Cory EN. Observations on the potato tuber moth. Journal of Economic Entomology. 1932;**25**:625-634

[123] Underhill GW. Studies on the potato tuber moth during the winter of 1925-1926. Virgina Experment Station Bulletin. 1926

[124] Langford GS. Winter survival of the potato tuber moth, *Phthorimaea operculella* Zeller. Journal of Economic Entomology. 1934;**27**:210-213

[125] Cory FM. The potato tuber moth. Vol. 57. University of Maryland Extension Service; 1925;**57**:3

[126] Langford GS. Observations on cultural practices for the control of the potato tuber-worm, *Phthorimaea operculella* (Zell.). Journal of Economic Entomology. 1933;**26**:135-137

[127] Rondon SI, SJ DB, Clough GH, Hamm PH, Jensen A. Ocurrence of the potato tuber moth in the Columbia Basin of Oregon and Washington. In: Kroschel J, Lacey, editors. Integrated Pest Management for the Potato Tuber Moth, *Phthorimaea operculella* Zeller - a Potato Pest of Global Importance. Tropical Agriculture 20, Advances in Crop Research 10. Weikersheim, Germany: Margraf Publishers; 2008. pp. 9-13

[128] Rondon SI, Murphy AF. Monitoring and controlling the beet leafhopper *Circulifer tenellus* in the Columbia Basin. American Journal of Potato Research. 2016;**93**(1):80-85. DOI: 10.1007/s12230-015-9491-3

[129] Klein M, Rondon SI, Walenta DL, Zeb Q, Murphy AF. Spatial and temporal dynamics of aphids (Hemiptera: Aphididae) in the Columbia Basin and Northeastern Oregon. Journal of Economic Entomology. 2017;**110**(4):1999-1910

[130] Broodryk SW. Ecological investigations on the potato tuber moth, *Phthorimaea operculella* (Zeller) (Lepidoptera: Gelechiidae). Phytophylactica. 1971;**3**:73-84

[131] Yathom S. Phenology of the potato tuber moth (*Phthorimaea operculella*), a pest of potatoes and processing tomatoes in Israel. Phytoparasitica. 1986;**17**:313-318

[132] Roelofs WL, Kochansky JP, Carde RT, Kennedy GG, Henrick CA, Labovitz JN, et al. Sex-pheromone of potato tuberworm moth, *Phthorimaea operculella*. Life Sciences 1975;**17**:699-706

[133] Persoons CJ, Voerman S, Verwiel PEJ, Ritter FJ, Nooyen WJ, Minks AK. Sex-pheromone of potato tuberworm moth, *Phthorimaea operculella* - isolation, identification and field evaluation. Entomologia Experimentalia Applicata. 1976;**20**:289-300

[134] Voerman S, Rothschild GHL. Synthesis of 2 components of sex-pheromone system of potato tuberworm moth, *Phthorimaea operculella* (Zeller) (Lepidoptera: Gelechiidae) and field experience with Them. Journal of Chemical Ecology 1978;**4**:531-542

[135] Raman KV. Control of potato tuber moth *Phthorimaea operculella* with sex pheromones in Peru. Agriculture, Ecosystems and Environment. 1988;**21**:85-99

[136] Lall L. Relationships between pheromone catches of adult moths, foliar larval populations and plant infestations by potato tuberworm in the field. Tropical Pest Management. 1989;**35**:157-159

[137] Horne PA. Sampling for the potato moth (*Phthorimaea operculella*) and its parasitoids. Australian Journal of Experimental Agriculture. 1993;**33**:91-96

[138] University-California. Integrated Pest Management for potatoes in the western United States. State wide integrated pest management program, Ag. and Nat. Resour. Pub. 3316, Western Regional Publication 011; 2006. 167 pp

[139] Rondon SI, Hervé M. Effect of planting depth and irrigation regimes on potato tuberworm (Lepidoptera: Gelechiidae) damage under central pivot irrigation in the Lower Columbia Basin. Journal of Economic Entomology. 2017;**110**(6):2483-2489

[140] Kennedy GG. Trap design and other factors influencing capture of male potato tuberworm (Lepidoptera: Gelechiidae) moths by virgin female baited traps. Journal of Economic Entomology. 1975;**68**:305-308

[141] Bacon OG, Seiber JN, Kennedy GG. Evaluation of survey trapping techniques for potato tuberworm moths (*Phthorimaea operculella*) with chemical baited traps. Journal of Economic Entomology. 1976;**69**:569-572

[142] Salas J, Alvarez C, Parra A. Evaluación de dos componentes de la feromona sexual, tres disenos y altura de colocación de trampas en la eficiencia de atracción y captura de adultos machos de *Phthorimaea opercuella*. Journal of Tropical Agronomy. 1991;**41**:169-178

[143] Tamhankar AJ, Harwalkar MR. Comparison of a dry and a water trap for monitoring potato tubermoth, *Phthorimaea operculella* Zeller. Entomology. 1994;**19**:163-165

[144] Clough GH, Rondon SI, DeBano SJ, David N, Hamm PB. Cultural practices to control the potato tuberworm. Journal of Economic Entomology. 2010;**103**(4):1306-1311

[145] Shorey HH, Deal AS, Hale RL, Snyder MJ. Control of potato tuberworms with Phosphamidon in southern California. Journal of Economic Entomology. 1967;**60**:892-893

[146] Bacon OG, McAlley NF, Riley WD, James RH. Insecticides for control of potato tuber worm and green peach aphid on potatoes in California. American Potato Journal. 1972;**49**:291-295

[147] Hofmaster RN, Waterfield RL. Insecticide control of the potato tuberworm in late-crop potato foliage. American Journal of Potato Research. 1972;**49**:383-390

[148] von Arx R, Goueder J, Cheikh M, Temime AB. Integrated control of potato tubermoth *Phthorimaea operculella* (Zeller) in Tunisia. International Journal of Tropical Insect Science. 1987;**8**:989-994

[149] BenSalah H, Aalbu R. Field use of granulosis virus to reduce initial storage infestation of the potato tuber moth, *Phthorimaea operculella* (Zeller), in North Africa. Agriculture, Ecosystems and Environment. 1992;**38**:119-126

[150] Ali MA. Effects of cultural practices on reducing field infestation of potato tuber moth (*Phthorimaea operculella*) and greening of tubers in Sudan. Journal of Agricultural Science. 1993;**121**:283-287

[151] Rahman A. Biology and control of volunteer potatoes: A review. New Zealand Journal of Expriemntal Agriculture. 1980;**8**(3):313-319

[152] Aarts HFM, Sijtsma R. Influence of crops on volunteer potatoes. In: Proceedings Crop Protection. Weeds; 1978. pp. 373-377

[153] Watmough RH, Broodryk SW, Annecke DP. The establishment of two imported parasitoids of potato tuber moth (*Phthorimaea operculella*) in South Africa: (*Copidosoma uruguayyensis, Apanteles subandinus*). Entomophaga. 1973;**18**:237-249

[154] Allemann J, Allemann A. Control and Management of Volunteer Potato Plants. South Africa: Univeristy of the Free State, Department of Soil, Crop, and Climate Sciences; 2013. pp. 46

[155] Clough G, DeBano S, Rondon SI, David N, Hamm PB. Use of cultural and chemical practices to reduce tuber damage from the potato tuberworm in the Columbia Basin. Hortscience. 2008;**43**:1159-1160

[156] von Arx R, Goueder J, Cheikh M, Bentemime A. Integrated control of potato tubermoth *Phthorimaea operculella* (Zeller) in Tunisia. Insect Science and Its Application. 1987;**8**:989-994

[157] Lloyd DC. Some South American parasites of the potato tuber moth, *Phthorimaea operculella* (Zeller) and remarks on those in other continents. Commonwealth Institute Biological Control Technichal Bulletin. 1972;**15**:35-49

[158] Howarth FG. Environmental impacts of classical biological control. Annual Review of Entomology. 1991;**36**:485-509

[159] Callan EM. Changing status of the parasites of potato tuber moth *Phthorimaea operculella* (Lepidoptera: Gelechiidae) in Australia. Entomophaga. 1974;**19**:97-101

[160] Meabed HAA, Rizk AM, El Hefnawy NN, El-Husseini MM. Biocontrol agents compared to chemical insecticide for controlling the potato tuber moth, *Phthorimaea operculella* (zeller) in the newly reclaimed land in Egypt. Egyptian Journal of Pest Control. 2011;**21**:97-100

[161] Broodryk SW. The biology of *Chelonus* (Microchelonus) *curvimaculatus* Cameron [Hymenoptera: Braconidae]. Journal of the Entomological Society of Southern Africa. 1969;**32**:169-189

[162] Baggen LR, Gurr GM. The Influence of Food on *Copidosoma koehleri* (Hymenoptera: Encyrtidae), and the use of flowering plants as a habitat management tool to enhance

biological control of potato moth, *Phthorimaea operculella* (Lepidoptera: Gelechiidae). Biological Control. 1998;**11**:9-17

[163] Morales J, Vásquez C, Pérez NL, Valera N, Ríos Y, Arrieche N, et al. Trichogramma species (Hymenoptera: Trichogrammatidae) parasitoids of lepidopteran eggs in Lara State, Venezuela. Neotropical Entomology. 2007;**36**:542-546

[164] Mandour NS, Mahmoud MF, Osman MAN, Qiu B. Efficiency, intrinsic competition and interspecific host discrimination of *Copidosoma desantisi* and *Trichogramma evanescens*, two parasitoids of *Phthorimaea operculella*. Biocontrol Science and Technology. 2008;**18**:903-912

[165] Berlinger MJ. Pests of processing tomatoes in Israel and suggested IPM model. Acta Horticulturae. 1992;**301**:185-192

[166] Berlinger MJ, Lebiush-Mordechi S. The potato tubermoth in Israel: A review of its phenology, behavior, methodology, parasites and control. Trends in Entomology. 1997; **1**:137-155

[167] Pucci C, Franco-Spanedda A, Minutoli E. Field study of parasitism caused by endemic parasitoids and by the exotic parasitoid *Copidosoma koehleri* on *Phthorimaea operculella* in central Italy. Bulletin of Insectology. 2003;**56**(2):221-224

[168] Kfir R. Fertility of the polyembryonic parasite *Copidosoma koehleri*, effect of humidities on life length and relative abundance as compared with that of *Apanteles subandinus* in potato tuber moth *Phthorimaea operculella*, biological control. The Annals of Applied Biology. 1981;**99**:225-230

[169] Choi JK, Kim JI, Kwon M, Lee JW. Description of the *Diadegma fenestrale* (Hymenoptera: Ichneumonidae: Campopleginae) attacking the potato tuber moth, *Phthorimaea operculella* (Lep.:Gelechiidae) new to Korea. Animal Systematics, Evolution and Diversity. 2013;**29**:70-73

[170] Aryal S, Jung C. IPM tactics of potato tuber moth *Phthorimaea operculella* (Zeller) (Lepidoptera: Gelechiidae): Literature Study. Korean Journal of Soil Zoology. 2015; **19**(1-2):42-51

[171] Ortu S, Flores I. Preliminart study on the control of *Phthorimaea operculella* on potato crops in Saridnia. Difesa delle Piante. 1989;**12**:81-88

[172] Odebiyi JA, Oatman ER. Biology of *Agathis gibbosa* (Hymenoptera: Braconidae), a primary parasite of the potato tuber worm. Annals of the Entomological Society of America. 1972;**65**:1104-1114

[173] Flanders RV, Oatman ER. Competitive interactions among endophagous parasitoids of potato tuberworm larvae in southern-California. Hilgardia. 1987;**55**:1-34

[174] Lloyd DC, Guido AS. Parasites of the potato tuber moth, *Gnorrimoschema operculella*. Commonwealth Institute of Biological Control Technical Bulletin. 1963;**3**:34

[175] Franzmann BA. Parasitism of *Phthorimaea operculella* (Lepidoptera: Gelechiidae) larvae in Queensland. Entomophaga. 1980;**25**:369-372

[176] Horne PA. The influence of introduced parasitoids on the potato moth, *Phthorimaea operculella* (Lepidoptera: Gelechiidae) in Victoria, Australia. Bulletin of Entomological Research. 1990;**80**:159-163

[177] Rao VP, Nagaraja H. Morphological differences between *Apanteles scutellaris* Muesebeck and *A. subandinus* Blanchard, parasites of the potato tubermoth, *Gnorimoschema operculella* (Zeller). Technichal Bulletin Commonwealth Institute of Control. 1968;**10**:57-65

[178] Mitchell BL. The biological control of potato tuber moth *Phthorimaea operculella* (Zeller) in Rhodesia. Rhodesia Agricultural Journal. 1978;**75**:55-58

[179] Divakar BJ, Pawar AD. Field recovery of *Chelonus blackburni* and *Bracon hebator* from potato tubermoth. Indian Journal of Plant Protection. 1979;**7**:214

[180] Leong JKL, Oatman ER. The biology of *Campoplex haywardi* (Hymenoptera: Ichneumonidae), a primary parasite of the potato tuberworm. Annals of the Entomological Society of America. 1968;**61**:26-36

[181] Choudhary R, Prasad T, Raj BT. Field evaluation of some exotic parasitoids of potato tubermoth, *Phthorimaea operculella* (Zeller). Indian Journal of Entomology. 1983;**45**:504-506

[182] Labeyrie V. Technique d'elevage de *Chelonus contratus* Nees, parasite de *Phthorimaea ocellatella* Boyd. Entomophaga. 1959;**4**:43-47

[183] Powers NR, Oatman ER. Biology and temperature responses of *Chelonus kellieae* and *Chelonus-phthorimaeae* (Hymenoptera: Braconidae) and their host, the potato tuberworm, *Phthorimaea operculella* (Lepidoptera: Gelechiidae). Hilgardia. 1984;**52**:1-32

[184] Keasar T, Steinberg S. Evaluation of the parasitoid *Copidosoma koehleri* for biological control of the potato tuber moth, *Phthorimaea operculella*, in Israeli potato fields. Biocontrol Science Technical Bulletin. 2008;**18**:325-336

[185] Cruickshank S, Ahmed F. Biological control of potato tuber moth, *Phthorimaea operculella* (Zell.) (Lepidoptera: Gelechiidae) in Zambia. Commonwealth Institute of Biological Control. Technichal Bulletin. 1973;**16**:62

[186] Kfir R. Biological control of the potato tuber moth Phth*orimaea operculella* in Africa. In: Neuenschwander P, Borgemeister C, Langewald J, editors. Biological Control in IPM systems in Africa. Oxfordshire, UK: CABI International Publishing; 2003. pp. 77-85

[187] Flanders RV, Oatman ER. Laboratory studies on the biology of *Orgilus-jenniae* (Hymenoptera, Braconidae), a parasitoid of the potato tuberworm, *Phthorimaea operculella* (Lepidoptera: Gelechiidae). Hilgardia. 1982;**50**:1-33

[188] Harwalkar MR, Rananavare HD, Rahaikar GW. Development of *Trichogramma brasiliensis* [Hym: Trichogrammatidae] on eggs of radiation sterilized females of potato tuberworm, *Phthorimaea operculella* [Lep.: Gelechiidae]. Entomophaga. 1987;**32**:159-162

[189] Zaki EN. Rearing of two predators, *Orius albidepennis* (Reut.) and *Orius laeviga-tus* (Fieber) (Hemiptera: Anthocoridae) on some insect larvae. Journal of Applied Entomology. 1989;**107**:107-109

[190] Montllor CB, Bernays EA, Cornelius ML. Responses of two hymenopteran predators to surface chemistry of their prey: Significance for an alkaloid-sequestering caterpillar. Journal of Chemical Ecology. 1991;**107**:107-109

[191] Horne PA, Edward CL, Kourmouzis T. *Dicranolaius bellulus* (Guerin-Meneville) (Coleop-tera: Melyridae: Malachiinae), a possible biological control agent of lepidopterous pests in inland Australia. Australian Journal of Entomology. 2000;**39**:47-48

[192] El-Sawi SA, Momen FM. Biology of some phytoseiid predators (Acari: Phytoseiidae) on eggs of *Phthorimaea operculella* and *Spodoptera littoralis* (Lepidoptera: Gelechiidae and Noctuidae). Acarologia. 2005;**45**:23-30

[193] Abd El-Gawad HAS, Sayed AMM, Ahmed SA. Functional response of *Chrysoperla carnea* (Stephens) (neuropteran: Chrysopidae) larvae to *Phthorimaea operculella* Zeller (Lepidoptera: Gelechiidae) eggs. Australian Journal of Basic and Applied Sciences. 2010;**4**:2182-2187

[194] Momen FM, El-Sawi SA. *Agistemus exsertus* (Acari: Stigmaeidae) predation on insects: Life history and feeding habits on three different insect eggs (Lepidoptera: Noctuidae and Gelechiidae). Acarologia. 2006;**46**:203-209

[195] NARC. Annual Report. Khumaltar, Lalitpur, Nepal: Nepal Agricultural Research Council. Entomology Division; 1995. 58 p

[196] Rusniarsyah L, Rauf A, Samsudin S. Pathogenicity and effectiveness of entomophatho-gen nematode Heterorhabditis sp. to potato tuber moth *Phthorimaea operculella* (Zeller) (Lepidoptera: Gelechiidae). Jurnal Silvikultur Tropika. 2015;**6**:1

[197] Salam KAA, Ghally SE, Kamel EG, Mohamed SA. Effect of gamma-irradiated ento-mopathogenic nematode *Steinernema carpocapsae* (Fil.) on the larvae of the potato tuber moth, *Phthorimaea operculella* (Zell.). Anzeiger Für Schädlingskunde Pflanzenschutz Umweltschutz. 1995;**68**:51-54

[198] Ivanova TS, Borovaya VP, Danilov LG. A biological method of controlling the potato moth. Zashchita Rastenii (Moskva). 1994;**2**:39

[199] Kakhki HM, Karimi J, Hosseini M. Efficacy of entomopathogenic nematodes against potato tuber moth, *Phthorimaea operculella* (Lepidoptera: Gelechidae) under laboratory conditions. Biocontrol Science and Technology. 2013;**23**:146-159

[200] Yuan HG, Lei ZR, Rondon SI, Gao YL. Potential of a strain of *Beauveria bassiana* (Hypocreales: Cordycipitaceae) for the control of the potato tuberworm, *Phthorimaea operculella* (Zeller). International Journal of Pest Management. 2017;**63**:1-3

[201] Yuan HG, Wu SY, Lei ZR, Rondon SI, Gao YL. Sub-lethal effects of *Beauveria bassiana* on field populations of the potato tuberworm, *Phthorimaea operculella* (Zeller) in China. Journal of Integrated Agriculture. 2018;**17**(4):911-918

[202] Yuan SY, Kong Q, Wang LJ, Wang JF, Yang JJ, Li XQ. Laboratory toxicity test on *Beauveria bassiana* MZ041016 to *Phthorimaea operculella* (Zeller). Jiangsu Agricultural Sciences. 2009;**6**:165-166

[203] Gao YL. Potato tuberworm: A threat for China potatoes. Entomology, Ornithology and Herpetology. 2018;**7**:e132

[204] Sewify GH, Abol-Ela S, Eldin MS. Effects of the entomopathogenic fungus *Metarhizium anisopliae* (Metsch.) and granulosis virus (GV) combinations on the potato tuber moth Phthorimaea operculella (Zeller) (Lepidoptera: Gelechiidae). Bulletin Faculty Agricultural University Cairo. 2000;**51**:95-106

[205] Steinhaus EA, Marsh GA. Previosuly unreported accessions for diagnosis and new records. Journal of Invertebrate Pathology. 1967;**9**:436-438

[206] Espinel-Correal C, Léry X, Villamizar L, Gómez J, Zeddam JL, Cotes AM, et al. Genetic and biological analysis of Colombian *Phthorimaea operculella* granulovirus isolated from *Tecia solanivora* (Lepidoptera: Gelechiidae). Applied and Environmental Microbiology. 2010;**76**:7617-7625

[207] Arthurs SP, Lacey LA, de la Rosa F. Evaluation of a granulovirus (PoGV) and *Bacillus thuringiensis* subsp *kurstaki* for control of the potato tuberworm (Lepidoptera: Gelechiidae) in stored tubers. Journal of Economic Entomology. 2008;**101**:1540-1546

[208] Foot MA. Susceptibility of twenty potato cultivars to the potato moth (*Phthorimaea operculella*) at Pukekohe: A preliminary assessment. New Zealand Journal of Experimental Agriculture. 1976;**4**:239-242

[209] Foot MA. Laboratory assessment of several insecticides against the potato tuber moth (*Phthorimaea operculella* Zeller). New Zealand Journal of Agricultural Research. 1976;**19**:117-125

[210] Chandramonhan N, Nanjan K. Damage level and control of potato tuber moth in Nilgiris district. The Madras Agricultural Journal. 1993;**80**:137-139

[211] Saour G. Effect of thiacloprid against the potato tuber moth, *Phthorimaea operculella* Zeller (Lepidoptera: Gelechiidae). Journal of Pest Science. 2008;**81**:3-8

[212] Mahdavi V, Rafiee-Dastjerdi H, Asadi A, Razmjou J, Fathi Achachlouei B, Kamita SG. Effective management of the *Phthorimaea operculella* using PVA nanofibers loaded with *Cinnamomum zeylanicum* essential oil. American Journal of Potato Research. 2017. DOI: 10.1007/s12230-017-9603-3

[213] Mahdavi V, Rafiee-Dastjerdi H, Asadi A, Razmjou J, Fathi Achachlouei B, Kamita SG. Synthesis of Zingiber officinale essential oil-loaded nanofiber and its evaluation on the potato tuber moth, *Phthorimae operculella* (Lepidoptera: Gelechiidae). Journal of Crop Protection. 2018;**7**(1):39-49

[214] Tanasković S, Djurović V, Popović B, Kneževic D, Gvozdenac S, Prvulović D. Plants as bio-insecticides in the service of the suppression of potato tuber moth in storage. XXI Savetovanje o Biotehncologijo. Zbornik radova. 2016;**21**(23)

[215] Moawad SS, Ami IM. Impact of some natural plant oils on some biological aspects of the potato tuber moth, *Phthorimaea operculella* (Zeller) (Lepidoptera: Gelechiidae). Egypt Agricultural and Biological Sciences. 2007;**3**:119-123

[216] Raman KV, Booth RH. Integrated control of potato moth in rustic potato storage. In: Proceedings of the Sixth Symposium on Inter. Sot. Trap. Lima, Peru: Root Crops International Potato center; 1984. pp. 509-515

[217] Raman KV, Booth RH, Palacios M. Control of potato tuber moth *Phthorimaea operculella* (Zeller) in rustic potato stores. Tropical Science. 1987;**27**:175-194

[218] Lal L. Studies on natural repellents against potato tuber moth *Phthorimaea operculella* in country stores. Potato Research. 1987;**30**(2):329-334

[219] Lal L. Potato tuber moth, *Phthorimaea operculella* (Zeller), in northeastern hills region and a simple method for its control. Indian Journal of Agricultural Science. 1988;**58**:130-132

[220] Niroula SP, Vaidya K. Efficacy of some botanicals against potato tuber moth. Our Nature. 2004;**2**:21-25

[221] Sharaby A, Abdel-Rahman H, Moawad S. Biological effects of some natural and echemical compounds on the potato tuber moth *Phthorimaea operculella* Zell. (Lepidoptera: Gelechiidae). Saudi Journal of Biological Sciences. 2009;**16**:1-0

[222] Tingey WM. Techniques for evaluating plant resistance to insects. In: Miller JR, Miller TA, Berenbaum M, editors. Insect Plant Interactions. New York, NY: Springer; 1986. pp. 251-284

[223] Panda N, Khush GS. Host Plant Resistance to Insects. Oxon, United Kingdom: CAB International; 1995

[224] Hardigan MA, Laimbeer PE, Newton L, Crisovan E, Hamilton JP, Vaillancourt B, et al. Genome diversity of tuber-bearing Solanum uncovers complex evolutionary history and targets of domestication in the cultivated potato. Proceedings of the National Academy of Sciences of the United States of America. 2017;**114**:E999-E10008

[225] Horgan FG, Quiring DT, Lagnaoui A, Pelletier Y. Tuber production, dormancy and resistance against *Phthorimaea operculella* (Zeller) in wild potato species. Journal of Applied Entomology. 2013;**137**:739-750

[226] Horgan FG, Quiring DT, Lagnaoui A, Salas A, Pelletier Y. Mechanism of resistance to tuber-feeding *Phthorimaea operculella* (Zeller) in two wild potato species. Entomologial Experiment Applied. 2007;**125**:249-258

[227] Das GP, Magallona ED, Raman KV, Adalla CB. Growth and development of the potato tuber moth, *Phthorimaea operculella* (Zeller), on resistant and susceptible potato genotypes in storage. Philippine Entomologist. 1993;**9**:15-27

[228] Malakar R, Tingey WM. Resistance of *Solanum berthaultii* foliage to potato tuberworm (Lepidoptera: Gelechiidae). Journal of Economic Entomology. 1999;**92**(2):497-502

[229] Golizadeh A, Razmjou J. Life table parameters of *Phthorimaea operculella* (Lepidoptera: Gelechiidae), feeding on tubers of six potato cultivars. Journal of Economic Entomology. 2010;**103**:966-972

[230] Douches DS, Li W, Zarka K, Coombs J, Pett W, Grafius E, et al. Development of Bt-cry5 insect-resistant potato lines 'Spunta-G2' and 'Spunta-G3'. Hortscience. 2002;**37**:1103-1107

[231] Douches DS, Pett W, Santos F, Coombs J, Grafius E, Metry EAWL, et al. Field and storage testing Bt potatoes for resistance to potato tuberworm (Lepidoptera : Gelechiidae). Journal of Economic Entomology. 2004;**97**:1425-1431

[232] Gatehouse AM, Norton R, Davinson E, Babbe SM, Newell CA, Gatehouse JA. Digestive proteolytic activity in larvae of tomato moth *Acanobia oleraceae*: Effects of plant protease inhibitors in vitro and in vivo. Journal of Insect Physiology. 1999;**45**(6):545-558

[233] Franco OL, Regden DJ, Melo FR, Grossi-de-Sa MF. Plant alpha amylase inhibiyors and their interaction with insect alpha amylases. European Journal of Biochemistry. 2002;**269**(2):397-412

[234] Fatehi S, RFP A, Bandani AR, Dastranj M. Effect of seed proteinaceous extract from two wheat cultivars against Phthorimaea oeprcuella digestive alpha-amylase and protease activities. Journal of the Entomological Research Society. 2016;**19**(1):71-80

[235] Saour G, Makee H. Radiation induced sterility in male potato tuber moth *Phthorimaea operculella* Zeller (Lepidoptera: Gelechiidae). Journal of Applied Entomology. 1997; **121**:411-415

[236] Saour G, Makee H. Effects of gamma irradiation used to inhibit potato sprouting on potato tuber moth eggs *Phthorimaea operculella* Zeller (Lep., Gelechiidae). Journal of Applied Entomology. 2002;**126**:315-319

[237] Saour G, Makee H. Susceptibility of potato tuber moth (Lepidoptera: Gelechiidae) to postharvest gamma irradiation. Journal of Economic Entomology. 2004;**97**:711-714

Susceptibility of Egg Stage of Potato Tuber Moth *Phthorimaea operculella* to Native Isolates of *Beauveria bassiana*

Nisreen Houssain Alsaoud,
Doummar Hashim Nammour and Ali Yaseen Ali

Additional information is available at the end of the chapter

http://dx.doi.org/10.5772/intechopen.78391

Abstract

The pathogenicity of three local isolates of the entomopathogenic fungus *Beauveria bassiana* (Bals.) Vuill was evaluated on eggs of potato tuber moth *P. operculella* (Zeller). The three isolates were coded as the following: B (isolate from Latakia), C (isolate from ICARDA) and D (isolate from Damascus). Three concentrations 10^4, 10^5, and 10^6, respectively, conidia/ml were used for each isolate. Eggs in the control were sprayed by sterilized water. All tests were done under laboratory conditions of temperature $28 \pm 2°C$ and relative humidity $40 \pm 5\%$. Susceptibility tests showed significant differences in averages of hatching rate between the control and both isolates B and C when 1×10^6 conidia/ml was applied, with averages 18.3 and 26.6% for previous isolates respectively, in contrast to 38.3 for isolate D and 66.6% for control. Findings indicated that eggs of *P. operculella* seemed sensible to local isolates of *B. bassiana* in varying degree, but further studies are required about the efficiency of effective isolates for controlling eggs of this pest in natural conditions.

Keywords: Beauveria bassiana, pathogencity, Phthorimaea operculella, Syrian isolates

1. Introduction

Potato tuber moth (Gelechiidae: Lepidoptera) *Phthorimaea operculella* is one of worldwide spread pests on potato and Solanaceae [1, 2]. The female lay her eggs on the leaves and non-covered tubers near to eyes (buds), larvae dig tunnels during their nutrition causes damages that reach approximately 100% on cultivated and stored potato [3, 4]. Therefore, this moth must be controlled in the field and in the store. There are many ways to control this pest starting by synthetic organic pesticides [5], natural origin insecticides like botanical extracts [6] and using genetically modified plants [5, 7, 8]. Natural parasitic enemies also successfully used like wasps from Braconidae, in addition to insect predators from Coccinellidae, Chrysopidae and Formicidae [9] and parasitic nematodes like *Steinernema carpocapsae*, *S. feltiae*, *S. glaseri*, and *Heterorhabditis bacteriophora* [10] which used successfully too. In last decade, biological origin insecticides like entomopathogenic viruses from group baculovirus [11] are used, as well as entomopathogenic fungi like *Beauveria bassiana* (Hypocreales: clavicipitaceae) [12, 13].

In this chapter, the pathogenicity of three native isolates of entomopathogenic fungus *Beauveria bassiana* was studied in different concentrations on eggs of potato tuber moth *Phthorimaea operculella* (Zeller), and it was determined the isolate which is the most pathogenicity on eggs, in vitro.

2. Materials and methods

2.1. Insect rearing

Tubers infected by potato tuber moth were collected from local markets, and laid on fine sand in aired glass cages (30 × 45 cm). Insects grew inside the cages in large numbers and fed by sugar solution (90%). Cages were supplied with fresh infected and noninfected tubers for activating insect rearing under laboratory conditions 28°C, R.H. 40 ± 5%.

2.2. Egg collecting

Eggs that are 1-day age were collected by using egg collecting chamber, which were prepared in the same manners of Maharjan [14], the chamber consists of plastic jar (7 × 15 cm) supplied with a cotton ball immersed in sugar solution. Five couples of moth adults (five males and five females) entered into the jar, and covered with gauze that was fixed by rubber, there is a piece of paper above the gauze. Eggs found on this paper were collected daily without opening the chamber [15]. Adults were fed by injecting the cotton ball (inside the jar) with sugar solution.

2.3. Infecting material and the spore suspension preparation

Three native isolates of the entomopathogenic fungus *B. bassiana* were used and were taken from different areas in Syria (**Table 1**).

Infecting material was prepared in safety bio-cabinet on Malt Extract Agar (MEA) medium in petri dishes of 9 cm in diameter, and dishes were placed inside a dark incubator at 25°C. Spores were harvested from dishes 2 weeks later, by adding 5 ml sterilized water for each dish then dish's contents were filtered across three layers of gauze. In addition, 5 ml of sterilized water was added over the gauze to assure collecting the maximum number of spores. The result liquid, which is considered as base solution 0.05% of tween 80%, was added to it.

Isolate name	Isolate code	Source	Isolate site
Latakia	B	Biological Enemies Center, Latakia	Soil of citrus orchard, Latakia
Icarda spt273	C	ICARDA, Aleppo	Isolated from dead *Eurygaster integriceps*, Aleppo
Damascus	D	Biotechnology Center, Damascus	Isolated from dead *Eurygaster integriceps*, Damascus

Table 1. Native isolates of entomopathogenic fungi *B. bassiana* and their sources and isolation sites.

Base solution concentration was determined by using a slide named Neubauer improved. The concentrations were adjusted for the three isolates to be: 10^4, 10^5, 10^6 conidia/ml. Control was treated with sterilized water and 0.05% of tween 80% was added to it.

2.4. Germination test

For testing the vitality of spores for each isolate, germination test was done in darkness under laboratory condition $28 \pm 2°C$, R.H. $40 \pm 5\%$ where 5 µl from each isolate, at the concentration of 10^4 spore/ml, was distributed on three drops on small petri dish of 5 cm diameter that contained Agar-Agar medium. Every drop represents a replicate that was covered by a covering glass before it was closed and then the dish was placed in a dark chamber. Next day, germinated spores were counted from 100 spores under every covering glass; after it was colored by lactophenol Cotton Blue, and the average of germination ratio was calculated for every isolate from its own dish.

2.5. Egg infecting by the fungus spore

A total of 600 eggs of potato tuber moth, 1 day age, were distributed into 30 carton cups that equal to 20 eggs/cup. Nine cups for each isolate distributed into three cups for each studied concentration (10^4, 10^5, 10^6 spore/ml), as well as three cups for control. Replicate eggs were sprayed with 2 ml of every solution by "Perfume Water Spray Bottle." Eggs in control were sprayed with 2 ml of sterilized water with 0.05% of tween 80% were added. After eggs spraying, inside their cups, they were covered with fine gauze which is fixed by rubber. Cups were placed in chamber at $28 \pm 2°C$, and R.H. $40 \pm 5\%$.

2.6. Readings

Hatching of treated eggs was observed to record the number of neonates, for 6 days period, in all treatments including the control. Hatched eggs ratio and dead eggs ratio were calculated when hatching was over, and nonhatched eggs were examined under 10× to record their color changes.

2.7. Data analysis

Percentage of corrected mortality was calculated according to Abbott [16].

$$\%\text{Corrected mortality} = (\%\text{mortality in control} - \%\text{mortality in treatment}) \times 100 / \tag{1}$$
$$(100\ \%\text{ mortality in control})$$

Data were analyzed by using SPSS program, where treatments were compared to test the significance of difference between averages by using LSD test at $p = 0.05$.

2.8. Scanning under electronic microscope

Eggs treated with local isolates of *B. bassiana* were observed under scanning electron microscope (SEM) and described in Science Faculty, Albaath University, according to its characteristics.

3. Results

3.1. Germination ratio

Averages of germination, after 24 h, were ranged between 47 and 67% (**Table 2**). Isolate B realized that has more germination ratio (67%) with significant difference from the isolate C which reached 48%, while isolate D reached to 55%. There is a significant difference in germination ratio between B and C.

3.2. Susceptibility of eggs

Reduction in hatchability rate was remarked in all treatments in comparison with control. Control realized a ratio of hatchability 66.6%, but this reduction in hatchability was not significant, between the control and the isolates, except in treatment with higher concentration (10^6 spore/ml) for the two isolates B and C (**Table 3**). Hatchability rate was 18.3% for isolate B and 26.6% for isolate C. There was no significant difference in hatchability between the isolate D and other treatments for the same previous concentration. Hatchability rates for the concentration 10^5 spore/ml were higher than 10^6 spore/ml, 33.3, 41.6 and 36.6% for isolates B, D and C, respectively.

Isolate name	Germination (average ± SE)
B	67 ± 5.77a
D	55 ±5.57ab
C	47 ± 4.33b
LSD	14.46

Means with same small letters in the same column have no significant differences at $p = 0.05$.

Table 2. Means of germination rates of native isolates of *B. bassiana* fungus at temperature $28 \pm 2°C$ and relative humidity $40 \pm 5\%$, 24 h after incubation.

Treatment-isolate/ concentration	Hatching rate of eggs (average ± SE)	Mortality of eggs (average ± SE)	Corrected mortality
Control	66.6 ± 14.5a	33.3 ± 14.5	—
Isolate B 10^6 conidia/ml	18.3 ± 1.6b	81.6 ± 1.6	72.5
Isolate D 10^6 conidia/ml	38.3 ± 13ab	61.6 ± 13	42.5
Isolate C 10^6 conidia/ml	26.6 ± 1.6b	73.3 ± 1.6	60
LSD value	25.84	—	—
Isolate B 10^5 conidia/ml	33.3 ± 10.9a	66.6 ± 10.9	50
Isolate D 10^5 conidia/ml	41.6 ± 6.6a	58.3 ± 11.5	37.5
Isolate C 10^5 conidia/ml	36.6 ± 6.6a	63.3 ± 6.6	45
LSD value	26.94	—	—
Isolate B 10^4 conidia/ml	35 ± 17.5a	65 ± 17.5	47.5
Isolate D 10^4 conidia/ml	45 ± 13.2a	55 ± 13.2	32.5
Isolate C 10^4 conidia/ml	43.3 ± 12.01a	56.6 ± 6.16	35
LSD value	38	—	—

Means with same small letters in the same column have no significant differences at $p = 0.05$.

Table 3. Means of hatching rates of *P. operculella* eggs after the treatment by different isolates of the entomopathogenic fungus *B. bassiana* with different concentrations at 28 ± 2°C and relative humidity 40 ± 5%.

Isolates in the concentration of 10^4 spore/ml were realized with higher hatchability rates: 35, 45, 43 and 67% for B, D, C and control; respectively.

In concentration 10^6 spore/ml, the isolate B reached the top corrected mortality rate on egg followed by C then D: 72.5, 60, 42.5%, respectively. These rates decreased to 50, 37.5 and 45% for the same previous isolates respectively, in concentration 10^5 spore/ml. In contrast, the lowest corrected mortality rates were in concentration 10^4 spore/ml: 47.5, 32.5 and 35% for B, D and C, respectively (**Table 2**).

3.3. Observing non-hatched egg

Number of dead eggs (nonhatched) resulted from the different treatments changed their color from transparent to yellowish or black. Black eggs percentage from dead eggs was 3, 7 and 20% for isolates D, C and B respectively in the concentration 10^6 spore/ml, as well as they were 8, 8 and 20% for the same previous isolates in the concentration 10^5 spore/ml and 5, 9 and 12% for the same isolates in the concentration 10^4 spore/ml (**Table 4**).

Eggs of potato tuber moth were shown under scanning electron microscopes (SEM) (**Image 1**), on the left, smooth egg's shell with some sculptures were shown under 400× as well as a slot of larva emergence at the pole of egg. On the right, infected egg by *B. bassiana* was seen under 800×, where spores have distributed on the surface especially in sculptures, but the density of spores was not so high where egg has hatched so it was shown a slot of larva emergence.

Treatment-isolate/ concentration	Transparent dead eggs % (average ± SE)	Yellow dead eggs % (average ± SE)	Black dead eggs (average ± SE)
Control	78 ± 8.82	19 ± 6.66	0 ± 0
Isolate B 10^6 conidia/ml	80 ± 12.58	0 ± 0	20.3 ± 12.02
Isolate D 10^6 conidia/ml	97 ± 13.23	0 ± 0	2.5 ± 1.66
Isolate C 10^6 conidia/ml	84 ± 1.66	16 ± 6	6.8 ± 2.88
Isolate B 10^5 conidia/ml	87 ± 12	5 ± 3.33	7.5 ± 2.89
Isolate D 10^5 conidia/ml	71 ± 4.4	8.5 ± 2.89	19.9 ± 4.4
Isolate C 10^5 conidia/ml	81 ± 8.82	10.4 ± 1.66	7.8 ± 2.88
Isolate B 10^4 conidia/ml	82 ± 10.93	12.8 ± 6	5 ± 1.66
Isolate D 10^4 conidia/ml	66.5 ± 9.28	21 ± 6	12 ± 6.66
Isolate C 10^4 conidia/ml	82 ± 7.26	8.8 ± 2.88	8.9 ± 2.88

Table 4. Means of coloration rates of dead *P. operculella* eggs after treatment with different isolates of the fungus *B. bassiana* with different concentrations at 28 ± 2°C and relative humidity 40 ± 5%.

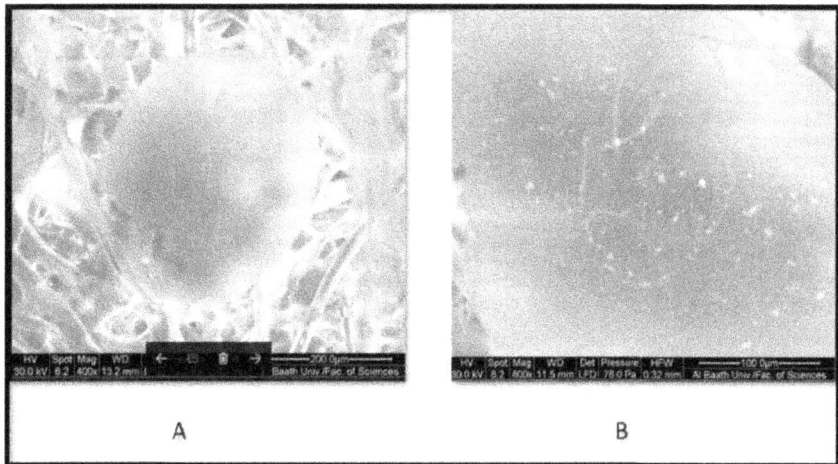

Image 1. Egg of *P. operculella* under SEM (A) noninfected under 400× (B) infected egg with *B. bassiana* under 800×.

4. Discussion

The results of this research showed the pathogenicity of the three native isolates of *B. bassiana* on egg stage of potato tuber moth, where they arrived at good death rates on eggs in different percentages between studied isolates. All results indicate to the ability of those isolates to infect the eggs stage of potato tuber moth, in spite of unsuitable condition of experiment especially the relative humidity (R.H.) was 40%, which was not optimum. That effected the germination of spores. In addition, the expression of virulence for those isolates affected negatively.

It is known that *B. bassiana* grows and germinates typically under 25–30°C and R.H. 100% [17]. Resulted death percentages from treatments with native isolates seem lower than results in similar study for the same fungus on the same moth, where egg death rate reached 76% after incubation at 25 ± 1°C and R.H. 80 ± 5% at the concentration 10^5 spore/ml of *B. bassiana* [12], while the rate did not exceed 67% in this research in the same concentration for the best isolate (B). That can be explained by the low relative humidity while this experiment.

On the other hand, the difference between the native isolates in pathogenicity on eggs, in this research may belong to the differences in vitality of spores and in germination rates. It is known that germinated spores have vitality and they can be active in control. Therefore, they penetrate the cuticle of insect [15, 18]. The results of germination rate showed that isolate B has higher rate of germination and with significant difference with D and C. However, it has the bigger chance to penetrate the host egg because the grand number of germinated spores on egg's surface. This issue may be the main reason for egg's infection was the greatest in B isolate in comparison with C and D isolates, that must has been a near percentage of infection depending on germination percentage.

In this study, the death of moth eggs may be to stop gas exchange between the egg and arounded air, where infected eggs by *B. bassiana* die and some of them became black because of fungal hyphae growth in the micropyles of egg shell [19]. Previous studies showed that eggs in most insects have sculptures that differ from one insect to another, where there is micropyle on front of egg pole which represents an entrance to sperm for fecundation operation. Also aeropyles aid in the exchange of oxygen and carbon dioxide and loss of some water. Woods [20] studied all those details delicately on eggs of *Manduca sexta* where he found that all aeropyles as well as the micropyle and egg shell in infected eggs were occupied by fungal hyphae. Therefore, gasses exchange operation decreased and the development of embryo died [20]. Therefore, infected eggs become sterile [19].

Shalaby et al. mentioned to coloration of *Tuta absoluta* eggs in black after its infection by B. bassiana. *T. absoluta* is Gelichiid [21] and black eggs resulted from infection by concentrations ranged between 10^7 and 10^{10} conidia/ml, death rate arrived 100%. Jaksch also mentioned [22] that eggs of *T. absoluta* showed spots in black as a result of direct infection with *B. bassiana*, eggs have dried clearly 4 days after incubation, white mycelium of fungus was observed on the pole of egg and a tissue of white spores have appeared on it.

Results of this research indicated that the most of nonhatching eggs had transparent color, and they represented nonfecunded eggs. Some of nonhatched eggs had yellowish color, and they represented fecunded eggs but they did not hatch for natural reasons, so they did not have a brown color as normal fecunded eggs before hatching [23–25]. The rest of dead eggs had a black color because of infection with *B. bassiana*, and it forms a percentage range from 0 to 20% of dead eggs. Low percentage of black eggs may belong to the low relative humidity and low concentrations in this research in comparison with another research on eggs of *Tuta absoluta*. Gottwald and Tedders [26] mentioned that decrease in fungus sporulation on dead host does not necessarily correlate to mortality that can be explained by several reasons like low temperature and relative humidity in its incubation climate or lose an essential substance for the development of fungus. For the same fungus, decrease in fungus sporulation on their dead hosts can be explained by diversity of the virulence between strains. That attributed to their genetic diversity that supports strains in its specialization in certain host and in its geographical distribution. [27].

Eggshell's structure has an important role in spore ability to adhere on the egg surface and increases the chance to infection impact. Therefore, egg's sensibility differs between species, for example, eggs of Lepidoptera have a huge chance of death as a result of fungal infection which belongs to sculptures on eggshell [20, 28]. *Ceratitis capitata* eggs are considered insensitive because they have smooth shell, where adhesion of spores is so difficult and the probability of their infection with entomopathogenic fungi seems relatively weak [29].

The importance of controlling eggs stage belongs to one hand, egg stage is fixed stage and easier in controlling than larvae in family Gelechiidae. On the other hand, Gelechiidae larvae (Ex: *P. operculella*, *T. absoluta* and *Scrobipalpa ocellatella*) dig tunnels inside leaves, tubers or roots, so they are protected from entomopathogenic effect. All previous makes its control so complex. Therefore, controlling eggs existing on leaves, fruits, stems and tubers represent a solution, where eggs are more exposed to natural enemies that prevent the appearance of damaging larvae from the beginning [22].

5. Conclusion

The pathogenicity of three local isolates of the entomopathogenic fungus *Beauveria bassiana* (Bals.) Vuill was evaluated on eggs of potato tuber moth *P. operculella* (Zeller). Isolates were taken from Latakia (isolate B), ICARDA spt273 (isolate C) and Damascus (isolate D). Three concentrations 10^4, 10^5, and 10^6 conidia/ml were used for each isolate; by spraying spore suspension on eggs. Eggs in the control were sprayed by sterilized water. The germination rate was evaluated after 24 h incubation in the dark. All tests were done under laboratory conditions of temperature $28 \pm 2°C$ and relative humidity $40 \pm 5\%$. Results showed significant differences in germination rate, where the average of germination rate was 67, 55, and 47% for isolates B, D and C respectively. Susceptibility tests showed significant differences in averages of hatching rate between the control and both isolates B and C when 1×10^6 conidia/ml was applied, with averages 18.3 and 26.6% for previous isolates respectively, in contrast 38.3% for isolate D and 66.6% for control. Findings indicated that eggs of *P. operculella* seemed sensible to local isolates of *B. bassiana* in varying degree. Results encourage further studies about the efficiency of effective isolates for controlling eggs of this pest in natural conditions of store and field and testing the local isolates on the other stages (adults and larvae) under better condition than this research condition.

Author details

Nisreen Houssain Alsaoud[1]*, Doummar Hashim Nammour[1] and Ali Yaseen Ali[2]

*Address all correspondence to: nisreensoud@gmail.com

1 Agriculture Faculty, Plant Protection Department, Albaath University, Homs, Syria

2 General Commission of Scientific Agricultural Research, Scientific Agricultural Research Center in Tarsus, Tarsus, Syria

References

[1] Ahmed AAI, Hashem MY, Mohamed SM, Khalil SHS. Protection of potato crop against *Phthorimaea operculella* (Zeller) infestation using frass extract of two noctuid insect pests under laboratory and storage simulation conditions. Archives of Phytopathology and Plant Protection. 2013;**46**(20):2409-2419. Available from: http://www.tandfonline.com/doi/abs/10.1080/03235408.2013

[2] Rondon SI. The potato tuber worm: A literature review of its biology, ecology, and control. American Journal of Potato Research. 2010;**87**(2):149-166

[3] Sporleder M, Zegarra O, Cauti EMR, Kroschel J. Effects of temperature on the activity and kinetics of the granulovirus infecting the potato tuber moth *Phthorimaea operculella* Zeller (Lepidoptera: Gelechiidae). Journal of Biological Control. 2008;**44**(3):286-295

[4] Visser D. Guide to Potato Pests and their Natural Enemies in South Africa. Pretoria: Arc-Roodeplaat Vegetable and Ornamental Plant Institute; 2005

[5] Dillard HR, Wicks TJ, Philp B. A grower survey of diseases, invertebrate pests, and pesticide use on potatoes grown in South Australia. Australian Journal of Experimental Agriculture. 1993;**33**(5):653-661

[6] Kroschel J. Management of the potato tuber moth *Phthorimaea operculella* Zeller (Lepidoptera, Gelechiidae)—An invasive pest of glob proportional. In: Proceedings of the Sixth World Potato Congress; August 20-26, 2006

[7] Arx RV, Gebhardt F. Effects of a granulosis virus, and *Bacillus thuringiensis* on life-table parameters of the potato tuber moth, *Phthorimaea operculella*. Journal of Entomophaga. 1990;**35**(1):151-159

[8] Cooper SG, Douches DS, Coombs JJ, Grafius EJ. Evaluation of natural and engineered resistance mechanisms in potato against Colorado potato beetle in a no-choice field study. Journal of Economic Entomology. 2007;**100**(2):573-579

[9] Coll M, Gavish S, Dori I. Population biology of the potato tuber moth, *Phthorimaea operculella* (Lepidoptera: Gelechiidae), in two potato cropping systems in Israel. Bulletin of Entomological Research. 2000;**90**(4):309-315

[10] Kakhki MH, Karimi J, Hosseini M. Efficacy of entomopathogenic nematodes against potato tuber moth, *Phthorimaea operculella* (Lepidoptera: Gelechiidae) under laboratory conditions. Biocontrol Science and Technology. 2013;**23**(2):146-159

[11] Carpio C, Dangles O, Dupas S, Léry X, López-Ferber M, Orbe K, Páez D, Rebaudo F, Santillán A, Yangari B, Zeddam JL. Development of a viral biopesticide for the control of the Guatemala potato tuber moth *Tecia solanivora*. Journal of Invertebrate Pathology. 2013;**112**(2):184-191

[12] Salih HAAM, Mahdi FA. The use of fungus *Beauveria bassiana* (Bulsamo) on tuber moth *Phthorimaea operculpella* (Zell.) in the Laboratory. Anbar Journal of Agricultural Sciences. 2010;**4**(8):334-341

[13] Hafez M, Zaki FN, Moursy A, Sabbour M. Biological effects of the entomopathogenic fungus, *Beauveria bassiana* on the potato tuber moth *Phthorimaea operculella* (Seller). Anzeiger für Schädlingskunde Pflanzenschutz Umweltschutz. 1997;**70**(8):158-159

[14] Maharjan R. Rearing methods of potato tuber moth, *Phthorimaea operculella* (Zeller) (Lepidoptera: Gelechiidae). Academia. 2011. http://www.academia.edu/2141878/Rearing_Methods_of_Potato_Tuber_Moth_Phthorimaea_operculella_Zeller_Lepidoptera_Gelechiidae

[15] Pekrul S, Grula EA. Mode of infection of the corn earworm (*Heliothis zea*) by *Beauveria bassiana* as revealed by scanning electron microscopy. Journal of Invertebrate Pathology. 1979;**247**(3):238-247

[16] Abbott WS. A method of computing the effectiveness of an insecticide. Journal of Economic Entomology. 1925;**18**(2):265-267

[17] Walstad JD, Anderson RF, Stambaugh WJ. Effects of environmental conditions on two species of muscardine fungi (*Beauveria bassiana* and *Metarrhizium anisopliae*). Journal of Invertebrate Pathology. 1970;**16**(2):221-226

[18] Boucias DG, Latgé JP. Nonspecific induction of germination of *Conidiobolus obscurus* and *Nomuraea rileyi* with host and non-host cuticle extracts. Journal of Invertebrate Pathology. 1988;**51**(2):168-171

[19] Barsagade DD, Pankule SD, Tembhare DB. Impact of fungus on egg shell of tropical tasar silk zorm, *Antheraea mylitta*: An ultra-structural approach. International Journal of Industrial Entomology. 2009;**18**(2):77-82

[20] Woods HA, Bonnecaze RT, Zrubek B. Oxygen and water flux across eggshells of *Mandoca sexta*. Journal of Experimental Biology. 2004;**208**:1297-1308

[21] Shalaby HH, Faragalla FH, El-Saadany HM, Ibrahim AA. Efficacy of three entomopathogenic agents for control the tomato borer, *Tuta absoluta* (Meyrick) (Lepidoptera: Gelechiidae). Journal of Nature and Science. 2013;**11**(7):63-27

[22] Jaksch. Selection of isolates of Entomopathogenic fungi for control of moth eggs. Akimoo; 2012 http://www.akimoo.com/selection-of-isolates-of-entomopathogenic-fungi-for-control-of-moth-eggs/

[23] Alvarez JM, Dotseth EJ, Nolte P. Potato tuber worm: A threat for Idaho potatoes. Moscow: University of Idaho Extension, Idaho Agricultural Experiment Station; 2005

[24] Rivera MJ. The potato tuber worm, *Phthorimaea operculella* (Zeller), in the tobacco, nicotiana. 2011

[25] Vaneva-Gancheva T. Morphological investigation on potato moth *phthorimaea operculella* zeller, lepidoptera, gelechiidae. Journal of Tabacco. 2009;**59**(3-4):81-87

[26] Gottwald TR, Tedders WL. Colonization, transmission, and longevity of *Beauveria bassiana* and *Metarhizium anisopliae* (Deuteromycotina: Hypomycetes) on pecan weevil larvae (Coleoptera: Curculionidae) in the soil. Environmental Entomology. 1984;**13**(2):557-560

[27] Coates BS, Hellmich RL, Lewis LC. Allelic variation of a *Beauveria bassiana* (Ascomycota: Hypocreales) minisatellite is independent of host range and geographic origin—Ge nome. Génome. 2002;**45**(1):125-132

[28] Arbogast RT, Leonard Lecato G, Van Byrd R. External morphology of some eggs of stored-product moths (Lepidoptera pyralidae, gelechiidae, tineidae). International Journal of Insect Morphology and Embryology. 1980;**9**(3):165-177

[29] Ali YA. Untersuchungen zur Effektivität entomopathogener Pilze im integrierten Pflan zenschutz am Beispiel der Fruchtfliegen *Ceratitis capitata* und *Rhagoletis cerasi* (Diptera, Tephriridae). 2010

Chapter 4

Moths of Economic Importance in the Maize and Sugar Beet Production

Renata Bažok, Zrinka Drmić, Maja Čačija,
Martina Mrganić, Helena Virić Gašparić and
Darija Lemić

Additional information is available at the end of the chapter

http://dx.doi.org/10.5772/intechopen.78658

Abstract

Maize and sugar beet productions are often threatened by various pests, causing high yield losses. Economically, most important maize pest is European corn borer, while sugar beet moth and noctuid moths cause serious damage on the sugar beet. This chapter highlights an introduction to several case studies representing long-term field research results on these pests. Depending on the pest, each study investigated the population level, dynamics of emergence or flight, damage levels and possibilities of forecasting on different localities in Croatia. The results could be of great importance in management of these pests. The European corn borer management depends mainly on timely conducted control, but the damage level also depends on maize hybrid and climatic conditions of investigated area. Damages caused by sugar beet moth depend on the population level and on locality's specific climate in a particular year. Sugar beet moth population and flight dynamics can be monitored by using pheromones, while pheromone application in forecasting and control showed to be disputable. Noctuid moths feed on the sugar beet foliage, causing high damages, especially on young plants. The damage level depends on the climatic conditions of the research area, and visual inspections of caterpillars are necessary for forecasting and control decision.

Keywords: maize, sugar beet, *Ostrinia nubilalis*, FAO maturity groups, *Scrobipalpa ocellatella*, flight dynamics, pheromones, forecasting, noctuid moths

1. Introduction

Maize is one of the most important field crops worldwide. In Europe, it is sown on almost 14 million of ha and in Croatia depending on the year, on between 250,000 and 300,000 ha [1].

Maize is usually attacked by a range of different pests, but the main pests in Europe are wireworms (family Elateridae), western corn rootworm (WCR) (*Diabrotica virgifera virgifera* LeConte) and European corn borer (ECB) (*Ostrinia nubilalis* (Hübner)).

The sugar beet is grown from subtropical areas to the northern regions of Scandinavia and originated near the Mediterranean Sea and Atlantic Ocean. Globally, 4.76 million hectares are sown with sugar beets every year, predominantly in the Russian Federation, Ukraine, USA, Germany, France, Turkey and Poland. The world's largest producer is France, with 32 million tons or 13.6% of total production [2]. In Croatia, the sugar beet has been sown since 1905. The area under sugar beet is approximately around 20,000 ha. In the past 3 years, the area sown by sugar beet has decreased up to 15,500 ha [3]. During the emergence of plants, sugar beets can be attacked by a large number of pests [4, 5]. Sugar beet seeds are treated with insecticides during seed processing, so in the early stages of germination and emergence, crops are protected from soil pests and some pests attacking young plants for 6 weeks if neonicotinoid insecticides are applied [6]. Sugar beet development extends through May, June, July and August. During vegetation, sugar beets are, due to favorable climatic conditions, increasingly attacked by a variety of pest species throughout Europe. Out of all species attacking sugar beet during this period, the species belonging to the order Lepidoptera are the most numerous. Čamprag [7] described 36 species from the order Lepidoptera that can cause serious damage to sugar beet crops. The most numerous family of harmful species is Noctuidae, which includes 29 species grouped in nine genera. Out of these 29 species, the most important species from the cutting species group are the black cutworm, *Agrotis ipsilon* Hufnagel, and the turnip moth, *Agrotis segetum* Denis & Schiffermüller [8]. From the surface-feeding species group, the most important are the cabbage moth *Mamestra brassicae* L., the bright line brown eye moth, *Lacanobia oleracea* L., and the silver Y moth, *Autographa gamma* L. [7, 9]. Besides the noctuid species, the beet moth, *Scrobipalpa ocelatella* Boyd (Lepidoptera: Gelechiidae) is shown to be growing problem in sugar beet production in neighboring countries (Serbia) [10] as well as in some years in Croatia [5].

The latest assessment by the United Nations Environment Programme and World Meteorological Organization-supported Intergovernmental Panel on Climate Change (IPCC), released in late 2014 concluded that climate change is already showing effects on many communities, with far greater impacts to come [11, 12]. The impacts of climate change on insect communities encompass changes impact on species life cycles [13, 14] or impacts on synchrony between host plant and herbivore [15].

The increasing pest population in particular region is very often correlated with climate change. This fact leads to the conclusion that the pest life cycle has to be investigated even though the data from the past already exist. This will allow us to record the changes in life cycle caused by climate change. These changes could result with increasing the importance of the particular pest.

1.1. European corn borer

The European corn borer (ECB) is the most important pest in Croatian agriculture [16, 17]. Maceljski [16] estimates the annual loss due to ECB to be 6–25%, while Ivezić and Raspudić [18] report average infestations of 50% during the period of 10 years. Despite significant damage, control of ECB has been attempted only in sweet corn and seed corn, while potential losses in commercial maize have not yet been addressed. Since sweet corn is meant for use in the fresh stage or for canning/freezing, control of ECB is absolutely necessary.

To achieve successful control of ECB, different alternative control methods have to be employed prior to the insecticide application. Agro technical methods (crop rotation, deep plowing, proper choice of sowing time), mechanical control and sowing resistant or tolerant varieties are very important. Since ECB larvae overwinter in the corn stalks, it is extremely important to mechanically destroy (cut) the corn stalks before the deep plowing in autumn. In Croatia, on some fields, corn stalks are left during the winter and this is enabling ECB larvae to successfully overwinter and increase the population level.

Resilience to the ECB is nowadays common with commercial maize hybrids. About 90% of 400 maize hybrids on market have shown a certain degree of resistance in vegetative phases of development [19]. Alongside resistance, modern maize hybrids are tolerant to a great degree to the damage caused by the ECB. Tolerance is the ability of a maize plant to withstand a certain population density of the insect without economic loss of yield or quality [19, 20]. The development of tolerant maize hybrids with a strong, robust stalk contributes immensely to reducing yield loss as a consequence of the damage caused by the ECB [21].

The main precondition for success in controlling ECB is correctly estimating the time when the insecticide should be applied [22]. According to Bažok et al. [22], the timing of ECB moth flights in Croatia has changed considerably since previous years. Bažok et al. [22] suggested using pheromones to demarcate the period of maximum moth incidence and to determine the percent of infested plants or the number of egg clusters on plant leaves by visual inspections carried out at the time of maximum incidence.

Therefore, the goal of research conducted in Croatia in 2017 was to establish the overwintering population level and dynamic of adult emergence of the European corn borer in North West Croatia. Additional goal was to estimate the differences between hybrids of different FAO maturity groups grown in areas with different climatic condition in terms of intensity of attack in vegetative maize growing stage (first ECB generation) and on maize cob (second ECB generation).

1.2. Moths on sugar beets

Surface-feeding noctuid moths are the most damaging pests in sugar beet, including the cabbage moth *Mamestra brassicae* L., the bright line brown eye moth, *Lacanobia oleracea* L., and the silver Y moth, *Autographa gamma* L. [7]. These species all have the potential to remove a majority (or all) of the above-ground foliage from young sugar beet plants, dramatically affecting plant growth and development. Cabbage and the bright line brown eye moth have two generations per year. They overwinter in pupae stage in the soil in the fields where the caterpillars lived. The butterfly eclosion starts at the end of May and early June. The second generation of adults flies in late July and early August. During the flying season, butterflies prefer planted areas for oviposition [23]. They lay eggs on sugar beet, but also on cabbages and other cultures and weeds. The first generation of caterpillar appears at the end of June and July, and the second generation appears at the end of August. The caterpillars are hygrophilic, preferring moisture areas in sugar beet crop. The maximum population level appears from the second half of June to the end of September. Silver Y moth is a migratory species partly developing population in our area, but most of the population comes from the southern regions. The butterfly eclosion in Croatia or arrival (from south) is similar to that of the previous two species, though the silver Y moth develops more generations per year (3–4), and it is possible that these generations overlap [23]. According to Čamprag [7], the economic thresholds established for the cabbage and the bright line brown eye moth are 10–12 caterpillars per m². For silver Y moth, precise threshold is not known, but 25% of the leaf area loss has been suggested as alternative economic threshold [24]. The population abundance and damages on sugar beet leaves are substantially impacted by air and soil temperature, as well as rainfall [7, 25–27]. Sugar beet seeds are treated with insecticides (often neonicotinoids) during seed processing and so, in the early stages of germination and emergence, crops are protected from pests for a short time [28]. Later in vegetation, sugar beets are increasingly attacked by a variety of pests as a result of favorable weather conditions. In this "unprotected" sugar beet period pheromone traps comprise one of the most effective methods for monitoring the seasonal flight dynamics of adult male moths [29–31]. These traps are often used to detect the presence of pests by season and location within a facility and to monitor apparent changes in the size of pest population over time [32]. However, the number of adult moths in pheromone traps is not always a direct indicator of the number of larvae, the life stage that damages the plants [33, 34]. Lemic et al. [28] established a strong positive correlation between captured male noctuid moths and the level of damage in sugar beet crop, and extreme relation of population density and weather conditions. However, for precise forecasting and decision about insecticide application in sugar beer field, visual inspections of moth damages are required [5, 7].

The sugar beet moth (*Scrobipalpa ocellatella* (Boyd) has been recorded in Croatia for the first time in 1947 in Slavonia region, and only 3 years later, it was mentioned as a pest present in almost all sugar beet growing area [35]. Sugar beet moth develops 4–5 generations in one vegetation season. It overwinters in last year sugar beer fields as adult caterpillar or in pupae stage [8, 16]. Since sugar beet moth has more generations and overwinters in different stages, often its generations overlap and all stages are present in sugar beets. For its reproduction, sugar beet moth prefers dry and warm weather, early spring and long autumn. In the sugar beet fields, Fajt [36] recorded that the attack starts at the field edges. Mines can be detected in the leaves and leaf stems, a distortion in the growing shoot with leaves spun tightly together are evidence of a larva within [35]. The danger increases in the second half of summer due to the increase in pest numbers in the second and subsequent generations. Economic threshold of damage: in the phenophase 6–8 leaves - 0.5 caterpillar per plant; at the beginning of the formation of root crops - 0.8–1 caterpillar per plant; at the beginning of the withering away of leaves - 2 caterpillars per plant. Sugar beet moth is a pest which appearance is irregular and systemic monitoring using pheromones allows occurrence detection in time [5].

The goal of the two studies carried out in Croatia in the period between 2012 and 2016 was to establish the dynamic of the flight and population level of sugar beet moth in Croatia and to establish the possibilities for beet moth forecasting by pheromone traps and visual inspections. Additionally, the goal was to establish the attack by various caterpillars (both moths and noctuids) in sugar beet fields planted in regions with different weather conditions.

2. Materials and methods

2.1. European corn borer

2.1.1. Study fields

Research was carried out in 2017 on four locations in Croatia: Šašinovečki Lug (45°51′00″ N, 16°10′01″ E), Vrana (43°56′45″ N, 15°26′53″ E), Tovarnik (45°13′28″ N, 19°21′38″ E) and Gola (46°1′44″ N, 16°33′13″ E). Besides these fields, at the location Šašinovečki Lug, eclosion and overwintering population of ECB was investigated on one unplugged maize field (45°51′21″ N, 16°10′17″ E).

2.1.2. Weather

Automatic weather stations were set up next to the cornfields on each location (Šašinovečki Lug, Vrana, Gola and Tovarnik), to collect data on average daily air temperature and daily amount of rainfall.

2.1.3. European corn borer moth eclosion and overwintering population

Samples of overwintering maize stalks (hybrid Bc 282) were collected on March 23, 2017. Twenty random stalks were collected from 15 rows. A total number of 300 maize plants were collected. The collected plants were 100 cm long. At the Department of Agricultural Zoology Faculty of Agriculture in Zagreb all plants were examined for shot holes from larval feeding. Maize stalks were cut into 20 cm pieces and placed in 15 entomological cages. Entomological cages were used for the purpose of rearing ECB larvae which overwintered in maize stalks. The eclosion of the moths out of the stalk was monitored every 7 days until May 29, 2017, when the final number and the gender of the enclosed moths were determined.

2.1.4. Estimating attack of first and second larval generation

In May 2017 on each of the four locations, 32 maize hybrids from four FAO maturity groups (300, 400, 500 and 600) were sown by permuted block randomization scheme. In total, on each location, 128 maize plots were planted, with each hybrid planted in four rows (4 m in length) and in four replications. The intensity of first ECB generation attack was estimated between June 28 and July 17, 2017. Damages on the plants (distinctive leaf holes, shot holes on stalks) were identified on two inner rows of every replication and recorded as the percent of the plants attacked by ECB. After all hybrids have been harvested (in September), 10 maize cobs were randomly selected from each replication and examined for second ECB generation damages and recorded as the percent of cobs infested by second ECB generation larva. For each FAO maturity group yield was recorded after harvesting.

2.1.5. Data analysis

In order to determine the difference in the intensity of the ECB attack (first and second ECB generation) among different FAO maturity groups, data on the percent of damaged maize plants and cobs on hybrids were submitted to two-way variance analysis (ANOVA). Averages were compared by Tukey's honestly significant range test. All differences were considered statistically significant at P = 0.05. Statistical evaluation of data was performed by the data management software ARM 9® GDM software, Revision 2018.2. [37].

2.2. Moths on sugar beets

Investigation of moths on sugar beets has been carried out in two separate studies.

2.2.1. Sugar beet moth

2.2.1.1. Study fields

To monitor the seasonal dynamics of sugar beet moth, a field trial was conducted in three growing seasons, from 2012 till 2014. Monitoring was performed from the early May to late August (18th to 35th week of the year) in Tovarnik (45°13′28″ N, 19°21′38″ E) in two sugar beet fields. In 2012, sugar beets were sown in fields of 45.89 ha and 4.82 ha; in 2013, sugar beets were sown in fields of 4.51 ha and 1.03 ha; and in 2014, sugar beets were sown in fields of 1.14 ha and 2.09 ha. The fields were approximately 5 km distanced.

2.2.1.2. Weather conditions

Prevailing weather conditions (i.e., mean air temperature and daily amount of precipitation) were collected from the two closest meteorological stations (Vukovar and Gradište) for all 3 years with the help of the Croatian Meteorological and Hydrological Service for each year of the period of investigation.

2.2.1.3. Monitoring of the moths and damage estimation

Pheromone traps (VARL + Csalomon®, Plant Protection Institute, Budapest, Hungary) were fixed on wooden sticks approximately 1.5 m above the ground and placed in the middle of the sugar beet fields. To catch the maximum number of specimens, the pheromone dispensers were changed at 6-week intervals as recommended by the manufacturer. Inspections on trapped moths were performed every 7 days.

Visual inspections of plants to detect damage caused by moth larvae were performed as described by Čamprag [23] in the Manual of the Reporting and Forecasting Service. Randomly, 10 × 10 sugar beet plants diagonally across the field were selected to detect damage on sugar beet leaves caused by larval feeding. In visual surveys, percentage of plants damaged by moths has been established as well as the number of caterpillars on the plants.

2.2.1.4. Data analysis

The moth monitoring results for the selected intervals are presented as the total number of males caught per trap per week. The average percent of damaged plants is presented as a function of the cumulative capture of moths in pheromone traps. Values were determined from the 18th until the 35th week of the year.

Data on moth abundance, percent of damaged plants and number of caterpillars on the plants, as well as meteorological data, were compared among years by ANOVA, and the mean separation was estimated using Tukey's HSD test [37]. The statistical software ARM 9® [37] was used to calculate correlation coefficients and to conduct regression analyses between the cumulative capture of male moths on pheromone traps and the percentage of plants damaged by larvae. The correlation coefficients were established, regression lines were described, and the coefficient of determination was calculated.

2.2.2. Noctuid moths

2.2.2.1. Study fields

The research was conducted during 2015 and 2016 on two distinct locations of Croatia, in Lukač (45°52′26″ N 17°25′09″ E) in Virovitica-Podravina county, and in Tovarnik (45°09′54″ N 19°09′08″ E) in Vukovar-Sirmium county. Untreated sugar beet (Artus in 2015 and Jelen in 2016) was sown at each location on 1000 m². In 2015, sowing was done on April 9 in Lukač and on April 11 in Tovarnik. In 2016, sowing was done few days earlier, that is, in Lukač on April 1 and on March 26 in Tovarnik.

2.2.2.2. Visual inspections and data analyses

Visual inspections of plants were performed as described by the Manual of the Reporting and Forecasting Service [23] to detect damage on leaves caused by noctuid larvae. Larval attack and damage on leaves was followed weekly on both locations in both years on randomly selected 4 rows of 20 m long starting form emergence of plants (i.e., May 6, 2015 and May 18, 2016) till root harvest (i.e., September 14, 2015 and September 7, 2016). Percentage of infected plants by moth larvae was recorded. Percentage of damage was calculated using the Townsend-Heuberger [38] formula. Data on percent of damaged plants as well as meteorological data were compared among years by ANOVA [37], and the mean separation was estimated using Tukey's HSD test. Where appropriate, data were $\sqrt{(x + 0.5)}$ transformed.

3. Results and discussion

3.1. European corn borer

In our survey, weather conditions during maize growing season varied among the investigated locations (**Table 1**). According to the data collected by weather stations, locality Vrana was characterized as having an extremely hot vegetation season. Locality Tovarnik was medium warm but had the lowest amount of rainfall (only 201 mm). By contrast, localities Šašinovečki Lug and Gola were characterized with high total amount of rainfall, especially locality Šašinovečki Lug which had more than 490 mm of rain. The weather conditions obviously could have an influence on the European corn borer population level and damages of first and second larval generation attack.

Locality	Average monthly temperature (°C)	Total amount of rainfall (mm)
Šašinovečki Lug	20.43	494.2
Tovarnik	21.12	201.6
Gola	20.00	399.5
Vrana	23.06	340.6

Table 1. Prevailing weather conditions during the vegetation season of maize in 2017 (May–September).

3.1.1. Overwintering population

After hibernation, ECB larvae developed into moths, whose eclosion out the stalk was monitored. The first enclosed moth was recorded on May 1, 2017. According to Kraljević Župić [39], the first moths were recorded on entomological lamps in location Sinj 31 day later, whereas in entomological cages with severed maize stalks they appeared somewhat earlier, but still more than 2 weeks later than in this research. The appearance of the first ECB generation in the field depends on the temperature and relative air moisture [16]. First eclosion according to Maceljski [16] usually takes place in the middle of May, although the majority of moths appear in June. Deviation in this research can be explained by the fact that the moths in cages were recorded as soon as they emerged from the cocoon, while several days must pass in order to catch the moths in a trap. Additionally, climatic condition influence the eclosion and as it has been found by Bažok et al. [22] in a very warm year, in 2003, the maximum of ECB moth appearance on pheromones on localities close to investigation site in this research, was in middle May. In total, 32 ECB moths developed from overwintering larvae. Male moths were the first to emerge out of the stalk (protandry). The total number of adult males was 14 (44%), whereas the total number of female moths was 18 (56%), which is in accordance with the research by Fadamiro and Baker [40] who also recorded a lower number of males compared to females. Considering these numbers and the fact that 32 moths developed from 300 stalks, it was estimated that in 1 hectare of unplugged maize cca. 8000 moths overwinter (at sowing 75,000 maize plants per hectare). This number of moths could produce more than 4 million larvae of first generation (estimated at cca. 500 eggs per female moth).

3.1.2. ECB first and second larval generation attack

Intensity of the first generation of ECB larval attack varied between 1.01% in Šašinovečki Lug to 38.1% in Tovarnik (**Table 2**). Significant differences in the intensity of the first generation of ECB larval attack was estimated among localities in all FAO maturity groups. High attack on all hybrids has been established in Tovarnik and Vrana, and lower attack has been established in Gola and the lowest in Šašinovečki Lug.

However, significant differences in maize stalk damage were estimated between FAO maturity groups in Šašinovec (condition of low attack) and in Tovarnik (conditions of high attack). In Vrana and Gola, no significant differences were established due to the high variability in attack intensity in different hybrids.

Locality	FAO maturity group				Tukey's HSD, P = 0.05
	300	400	500	600	
Šašinovečki Lug	1.18d'B**	1.20 dAB	3.05 bA	1.45 dAB	2.24
Tovarnik	25.28 aB	27.59 aAB	31.57 aAB	38.1 aA	11.47
Gola	5.6 c	7.54 c	7.48 b	9.5 c	ns
Vrana	15.94 b	17.56 b	23.9 a	24.71 b	ns
Tukey's HSD, P = 0.05	3.36	3.69	5.07	4.89	/

*Means followed by the same small letter do not significantly differ among localities (i.e., columns) (P < 0.05; Tukey's honestly significant difference (HSD)).

**Means followed by the same capital letter do not significantly differ among FAO maturity groups (i.e., rows) (P < 0.05; Tukey's honestly significant difference (HSD)).

Table 2. Intensity of the first-generation ECB larval attack (%).

The intensity of the attack of the second generation of ECB was much higher comparing to the first generation, it ranged between 17.19 and 92.81% (**Table 3**). The differences among localities in the attack of each FAO maturity groups were significant. The highest attack on all FAO maturity groups has been established in Tovarnik, somewhat lower in Vrana moderate in Šašinovečki Lug and the lowest in Gola. This situation is very similar with those established for the first generation. The localities with higher attack of the first generation, Tovarnik and Vrana, had higher attack of the second generation too. The locality Šašinovečki Lug had the lowest attack of the first generation, but the attack of the second generation was moderate and higher comparing to the locality Gola. This is probably the consequence of the higher amount of rainfall received in July in Šašinovečki Lug comparing to Gola.

The differences among the FAO maturity groups were significant at two localities, Šašinovečki Lug and Gola where the attack was low to moderate. In described conditions, FAO maturity

Locality	FAO maturity group				Tukey's HSD, P = 0.05
	300	400	500	600	
Šašinovečki Lug	1.18d'B**	1.20 dAB	3.05 bA	1.45 dAB	2.24
Tovarnik	25.28 aB	27.59 aAB	31.57 aAB	38.1 aA	11.47
Gola	5.6 c	7.54 c	7.48 b	9.5 c	ns
Vrana	15.94 b	17.56 b	23.9 a	24.71 b	ns
Tukey's HSD, P = 0.05	3.36	3.69	5.07	4.89	/

*Means followed by the same small letter do not significantly differ among localities (i.e., columns) (P < 0.05; Tukey's honestly significant difference (HSD)).

**Means followed by the same capital letter do not significantly differ among FAO maturity groups (i.e., rows) (P < 0.05; Tukey's honestly significant difference (HSD)).

Table 3. Intensity of the second-generation ECB larval attack (%).

groups 400 and 500 had higher attack comparing to FAO groups 300 and 600. According to available literature, the average intensity of ECB attack in Croatia varies and depends on the FAO maturity group, locality and year of investigation. Some authors have shown that with the increase of the length of vegetation period (i.e., maturity group) the intensity of the attack of ECB is increasing [17]. Our investigation cannot prove these findings because the FAO group 400 had the highest attack on the sites where the overall attack was lower (i.e., Šašinovečki Lug and Gola). The same time, on the sites where the overall attack of ECB was high (i.e., Tovarnik and Vrana) the attack of FAO 400 was high. This is in line with the results reported by Augustinović et al. [41] who reported the same time the lowest and the highest attack of FAO 400, depending on the site of investigation. Investigations conducted by Raspudić et al. [42] have not shown statistically relevant differences in attack intensity among the hybrids what is similar for our results obtained at the localities Tovarnik and Vrana.

Maize yield was recorded for each FAO maturity group on all localities and standardized (14% moisture) and is presented in **Table 4**. There is a great difference between locations, which was implied regarding the weather conditions, but no significant differences in yield between FAO maturity groups were found.

Previous research conducted on yield in maize hybrids of different FAO maturity groups shown that the significantly highest yield should be expected for the FAO 500 and 600 maturity groups [43–45]. It can be assumed that significantly higher ECB attacks (both generations) on medium-late FAO maturity groups have resulted in yield reduction and the yield was lower than expected for these FAO groups and did not differ from the yield of early to medium FAO maturity groups.

3.2. Moths on sugar beets

3.2.1. Sugar beet moth

In our survey, weather conditions during growing season varied among the investigated years (**Table 5**). According to the Croatian Meteorological and Hydrological Service in 2012, Croatia was characterized as having an extremely hot and dry year. By contrast, 2013 was characterized as a moderate year with average air temperatures, medium amount of total amount of rainfall, and 2014 was characterized as cold and moist. In 2012, the investigated

Locality	FAO maturity group			
	300	400	500	600
Šašinovečki Lug	7.3	7.5	7.6	8.2
Tovarnik	5.2	5.2	5.1	5.2
Gola	25.3	26.6	26.1	25.1
Vrana	14.6	16.9	14.0	15.0

Table 4. Maize field yield (t/ha) on four localities in 2017.

Year	Mean air temperature (°C) ± SD	Mean soil temperature (°C) ± SD	Total amount of rainfall (mm) ± SD
2012	22.26 ± 0.03 a*	24.75 ± 0.26 a	144.71 ± 30.26 b
2013	21.06 ± 0.08 b	23.59 ± 0.45 b	272.75 ± 31.32 ab
2014	19.99 ± 0.19 c	23.05 ± 0.5 b	400.50 ± 3.39 a
HSD P = 0.05	0.646	0.74	181.626

*Values followed by the same letter are not significantly different among columns (P > 0.05; Tukey's HSD).

Table 5. Characteristics of the climatic conditions prevailing in the years of investigation (from 18th till 35th week).

period was characterized by higher mean air temperatures (22°C) and mean soil temperatures at 10 cm depth (25°C). Consequently, the total amount of rainfall in the same period was significantly lower (144 mm), while 2013 was characterized as a moderate year with average air temperatures of 21°C and average soil temperatures of 24°C, with a total amount of rainfall of 272 mm. The investigation period in 2014 was characterized by lower mean air temperatures (19°C), lower soil temperatures (23°C) and a significantly higher amount of rainfall (400 mm) compared to 2012 and 2013. The weather conditions evidently had a great influence on the male moth population level and population dynamics.

The presence of sugar beet moth has been established through the whole vegetation period in 2012 and in 2014, while in 2013, the moths were captured on the traps until week 26 (**Figure 1**). Later on, male moth captures have not been observed. The reason could be found in the very high prevailing temperatures in the whole second part of the vegetation season in 2013. During the period of 8 weeks in 2013 (from mid of July to mid of September), the average weekly temperatures were over 24°C, and the amount of rainfall was extremely low. In spite of the fact that sugar beet moth prefers dry and hot years for its reproduction [10], the prevailing conditions in 2013 were not very favorable for moth development. It is difficult to state how many generations sugar beet moth developed in 1 year. There are four peaks of the flight in 2012 and 2014, but the number of moths caught was too low to

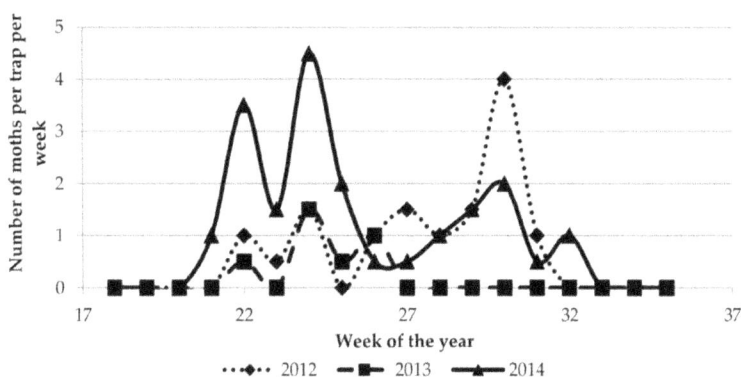

Figure 1. Total moths capture per pheromone trap per week (from 18th till 35th) in Tovarnik, Croatia, 2012–2014.

conclude that the moth developed four generations as it is stated by Maceljski [16] and Čamprag et al. [10]. Based on our results, which confirm previous surveys of Čamprag et al. [10], by the use of pheromones, we can predict the abundance of the first generation of moth which happens in 21st and 22nd week of the year. In subsequent years, the flight dynamics has similar patterns in the first 6 weeks of the moth appearance. In the second part of the vegetation season (week 27 till week 35), flight dynamic patterns depended very much on the prevailing weather conditions.

Sugar beet moth can cause important damage during the vegetation season of sugar beets. The surface-feeding larvae are foliage-feeding pests, but their later generation enters into the sugar beet root so they can be extremely harmful due to the destruction of leaf mass as well as due to the damaging the sugar beet roots and opening the floor to the infections with different pathogens [7, 16], which has a negative effect on sugar accumulation in the root [43]. Thus, possible damage forecasting and thresholds for suppression based on male moth captures on pheromone traps can be of great importance in the management of sugar beet pests.

We observed a correlation between male moth captures and plant damage in all investigated years in spite of the differences in weather conditions, which directly caused differences in population dynamics and differences in the total capture of moths on pheromones (**Figure 2**). The correlation coefficients were high for 2012 and 2013 and could be described as full positive correlations and medium in 2014 and could be described as positive correlation [46]. The coefficients of determination (r^2) were also high for both species groups, and the regression curves had similar tendencies and were linear (**Table 6**).

Moth population growth during the vegetation period increased the damage to sugar beet plants. In warm and dry years (e.g., 2012 and 2013), 10 collected sugar beet male moths caused

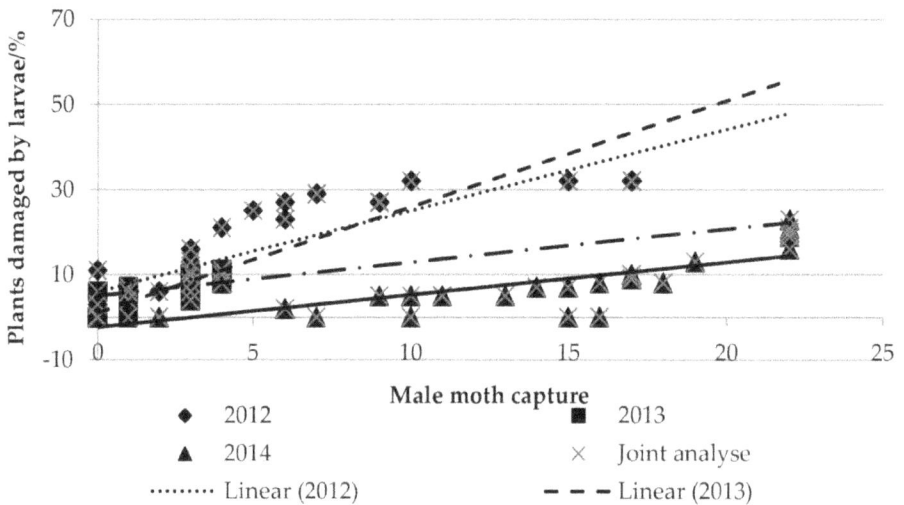

Figure 2. Regression analysis of the cumulative capture of sugar beet male moths on pheromone traps (x) versus the percentage of plants damaged by moth larva (y), Tovarnik, Croatia, 2012–2014.

Year	n	Correlation coefficient (r)	Coefficient of determination (r²)	Probability (p)	Regression equation
2012	36	0.7817	0.8841	0.0001	$y = 1.0959x + 5.9892$
2013	36	0.8283	0.9101	0.0001	$y = 2.4935x + 2.0284$
2014	36	0.6711	0.8192	0.0001	$y = 0.762x - 2.3334$
Joint analysis	108	0.2838	0.5327	0.0001	$y = 0.7835x + 5.0267$

Table 6. Correlation coefficients, coefficients of determination and regression equations for percent of plants damaged by sugar beet moth larvae (y) and cumulative capture male sugar beet moths on pheromone traps (x), Tovarnik, Croatia.

the damage on 25–30% plants. However, in the year in which weather conditions were not so hot and dry (e.g., 2014), the same number of male moths indicated the damage of 5% of damaged plants. In 2013, we did not record neither new male moth capture nor the additional damage on the plants, and analysis is based on the data from first part of the season when the climatic conditions were preferable for moth development. Later on, the climatic conditions were not favorable and moth population reduced so the larvae did not continue to cause the damage on the plants.

Although we established a strong correlation between the cumulative number of male moths caught on pheromone traps and damage on plants, we were not able to detect a threshold for decisive control because we used sex pheromone-baited traps in our investigations, which, while highly sensitive and selective, have the inherent weakness of attracting only male moths. Therefore, traps that attract female moths would potentially provide more valuable information for pest control decisions.

3.2.2. Noctuid moths

On locations Lukač and Tovarnik weather conditions during sugar beet growing season varied among the investigated years and locations (**Figure 3**). On both locations, year 2015 was characterized with higher air and soil temperatures and lower amount of precipitation comparing to the year 2016. In both investigation years, location Tovarnik was characterized by higher mean air temperatures comparing to location Lukač. Consequently, the total amount of rainfall in the same period was significantly lower (higher in Lukač). The weather conditions evidently had a great influence on the noctuid larval attack on sugar beet.

3.2.2.1. Visual inspections of leaf damage

The attack of harmful caterpillars was determined throughout the vegetation in both research years. A total of 22 visual inspections were performed (depending on the year). The percentage of caterpillar-damaged plants is shown in **Figure 4**. Although the plants were found to be damaged, caterpillars have been rarely found. In 2015, the maximum infestation was 0.45 caterpillars per plant, which is below the threshold. In 2016, the maximum infestation of caterpillars was even lower.

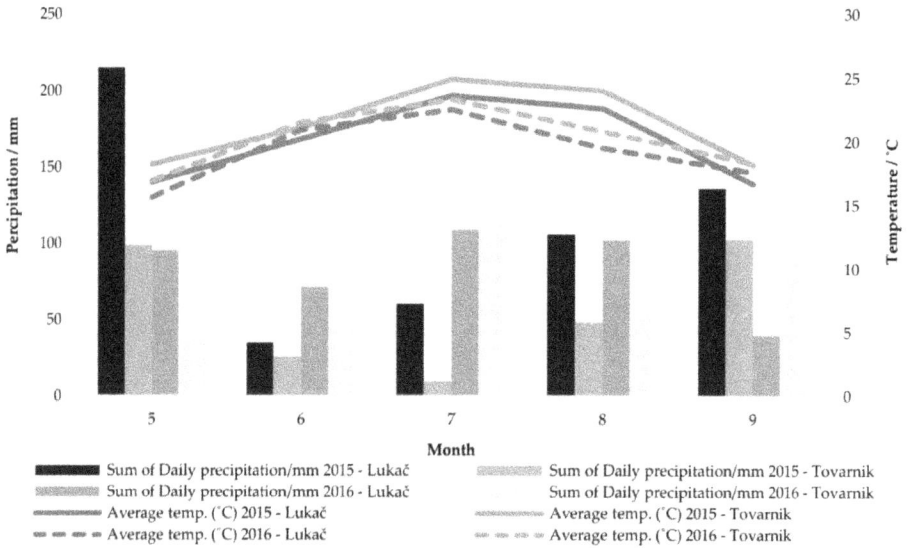

Figure 3. Weather conditions in two locations where research was conducted.

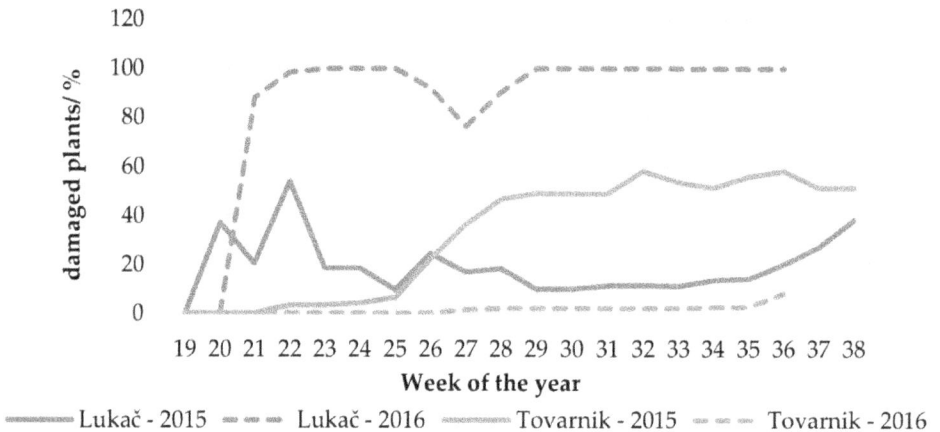

Figure 4. Noctuid moth larvae damage dynamics in sugar beet.

A warm summer with low humidity preceded higher egg mortality and a second generation of larvae in low numbers [47]. Indeed, damages on leaves from the second-generation larvae were not significantly higher. Warm and dry conditions in Tovarnik had a negative influence on the first generation of noctuid moth larvae, which directly caused lower damage dynamic in Tovarnik versus Lukač in whole investigation period. In 2015, the larval damage on sugar beet on both locations was lower. However, weather conditions in 2015 were favorable for noctuid moth development (lower temperature, higher precipitation; [28]) and, in 2016, a population recovery was observed which was visible from higher level of larval damage. These results confirm a previous survey by Vajgand [27] in which a decrease in the moth population and larval damages were caused by a very warm and dry vegetation period.

4. Conclusions

The results of the investigation could be of great importance in management of investigated pests, ECB and moths (sugar beet moth and noctuid moths) on sugar beet.

4.1. European corn borer

In North West Croatia, the eclosion of the European corn borer overwintering population monitored in cages happened about 2 weeks earlier (beginning of May) than previously recorded in the literature. Male moths emerged first (protandry), and in total population they were represented in lower numbers than female moths. Changes in timing of ECB moth flight and, consequently, changes in the period of maximum moth incidence have a great influence on the success of ECB control, as the insecticides must be applied in timely manner. Also, the intensity of the first and second ECB larval generation attack varied significantly among four FAO maturity groups and among four investigation sites in all FAO maturity groups, the latest presumably due to different weather conditions. Significant differences in maize stalk damage, caused by the first generation, were recorded between FAO maturity groups on two locations which differed in attack intensity. Similarly, the second-generation attack differed significantly among FAO maturity groups on two locations where the attack was low to moderate. Results confirmed that the damage of ECB is determined by the weather conditions rather than by FAO maturity group.

4.2. Moths on sugar beets

The seasonal dynamics for sugar beet moth has shown that, during the 3-year period, it appeared between 21st and 22nd week of the year, suggesting the pheromones could be used to predict the first generations abundance. After 27th week, the flight dynamics depended on the prevailing climatic conditions. Four peaks of flight were detected, but due to low moth number, we cannot conclude on number of generations per year. A strong correlation between male moth captures and plant damage suggested that moth population growth increased the damage on sugar beet. However, the same number of male moths did not cause the same level of damage in years with different climatic conditions. Given the fact we used sex pheromones, which attracted only moth males, we were not able to conclude on a threshold for decisive control; therefore, pheromones which also attract females could be useful in sugar beet moth forecasting and control decisions. Noctuid moth damages on sugar beet leaves, determined by visual plant inspections, showed that the damages depended on climatic conditions of the location and decreased in very warm and dry conditions.

Acknowledgements

This chapter was supported by Croatian Science Foundation project: "09/23 Technology transfer in sugar beet production: improvements in pest control following the principles of integrated pest management (IPM)", IPA grant number 2007/HR/16IPO/001-040511 "Enhancement of collaboration between science, industry and farmers: Technology transfer for integrated pest management (IPM) in sugar beet as the way to improve farmer's income and reduce pesticide use," the European Union from the European Social Fund within the project "Improving human capital by professional development through the research program in Plant Medicine" (HR.3.2.01-0071) and partly supported by the Environmental Protection and Energy Efficiency Fund and Croatian Science Foundation trough project: AGRO-DROUGHT-ADAPT 2016-2106-8290 "Adaptability assessment of maize and soybean cultivars of Croatia in the function of breeding for drought tolerance."

Author details

Renata Bažok*, Zrinka Drmić, Maja Čačija, Martina Mrganić, Helena Virić Gašparić and Darija Lemić

*Address all correspondence to: rbazok@agr.hr

Faculty of Agriculture, University of Zagreb, Zagreb, Croatia

References

[1] FAO. World Agricultural Production [Internet]. 2018. Available from: http://www.fao. org/faostat/en/#home [Accessed: Apr 20, 2018]

[2] Pospišil M. Ratarstvo: II.dio – industrijsko bilje. Čakovec: Zrinski; 2013

[3] Central Bureau of Statistics. Statistical Yearbook Republic of Croatia [Internet]. Croatia: Croatian Bureau of Statistics; 2017. Available from: https://www.dzs.hr/hrv/publication/ stat_year.htm [Accessed: Apr 4, 2018]

[4] Čamprag D. Štetočine i bolesti šećerne repe i njihovo suzbijanje. Zavod za selekciju šećerne repe-Crvenka. 1954

[5] Bažok R. Suzbijanje štetnika u proizvodnji šećerne repe. Glasilo biljne zaštite. 2010; **10**(3):153-165

[6] Igrc Barčić J, Dobrinčić R, Šarec V, Kristek A. Istraživanje tretiranja sjemena šećerne repe insekticidima. Poljoprivredna znanstvena smotra. 2000;**65**(2):89-97

[7] Čamprag D. Štetočine šećerne repe u Jugoslaviji, Mađarskoj, Rumuniji i Bugarskoj sa posebnim osvrtom na važnije štetne vrste. Novi Sad, Serbia: Poljoprivredni fakultet; 1973

[8] Čamprag D, Jovanić M. Cutworms (Lepidoptera: Noctuidae)-Pests of Agricultural Crops. Novi Sad, Serbia: Poljoprivredni fakultet; 2005

[9] Metspalu L, Jogar K, Hiiesaar K, Grishakova M. Food plant preference of the cabbage moth *Mamestra brassicae* (L.). Latvian Journal of Agronomy. 2004;**7**:15-19

[10] Čamprag D, Sekulić R, Kereši T. Repina korenova vaš (*Pemphigus fuscicornis* Koch) s posebnim osvrtom na integralnu zaštitu šećerne repe od najvažnijih štetočina. Novi Sad, Serbia: Poljoprivredni fakultet; 2003

[11] IPCC – Intergovernmental Panel on Climate Change. Climate Change 2014: Synthesis Report. In: Core Writing Team, Pachauri RK, Meyer LA, editors. Contribution of Working Groups I, II and III to the Fifth Assessment Report of the Intergovernmental Panel on Climate Change [Internet]. Geneva, Switzerland: IPCC; 2014. p. 151. Available from: http://www.ipcc.ch/pdf/assessmentreport/ar5/syr/SYR_AR5_FINAL_full_wcover.pdf [Accessed: Feb 23, 2016]

[12] UNEP – United Nations Environment Programme. Minimizing the Scale and Impact of Climate Change [Internet]. 2015. Available from: http://www.unep.org/annualreport/2014/en/pdf/climate_change.pdf [Accessed: Feb 23, 2016]

[13] Musolin DL. Insects in a warmer world: Ecological, physiological and life-history responses of true bugs (Heteroptera) to climate change. Global Change Biology. 2007;**3**(8):1565-1585. DOI: 10.1111/ j.1365-2486.2007.01395.x

[14] Kocmánková E, Trnka M, Eitzinger J, Dubrovský M, Štěpánek P, Semerádová D, Balek J, Skalák P, Farda A, Juroch J, Žalud Z. Estimating the impact of climate change on the occurrence of selected pests at a high spatial resolution: A novel approach. Journal of Agricultural Science. 2011;**149**:185-195. DOI: 10.1017/S0021859610001140

[15] Junk J, Eickermann K, Görgen K, Beyer M, Hoffmann L. Ensemble-based analysis of regional climate change effects on the cabbage stem weevil (*Ceutorhynchus pallidactylus* (Mrsh.) in winter oilseed rape (*Brassica napus* L.). Journal of Agricultural Science. 2012;**150**:191-202. DOI: 10.1017/S0021859611000529

[16] Maceljski M. Poljoprivredna entomologija. Čakovec: Zrinski; 2002

[17] Ivezić M, Raspudić E. Intensity of attack of the corn borer (*Ostrinia nubilalis* Hubner) on the territory of Baranja in the period 1971-1990. Natura Croatica 1997;**6**:137-142

[18] Ivezić M, Raspudić E. Ekonomski značajni štetnici kukuruza na području istočne Hrvatske. Razprave IV. Razreda SAZU [dissertationes]. 2005;**XLV-1**:87-98

[19] Barry BD, Darrah LL. Impact of mechanisms of resistance on European corn borer in selected maize hybrids. In: Mihm JA, editor. Insect Resistant Maize: Recent Advances and Utilization. Proc. Int. Symp. on Methodologies for Developing Host Plant Resistance to Maize Insects, CIMMYT, Mexico DF. 27 Nov-3 Dec. 1994:25-32

[20] Guthrie WD, Barry BD. Methodologies used for screening and determining resistance in maize to the European corn borer. In: Towards insect resistant maize for the third world. Proc. Int. Symp. on Methodologies for Developing Host Plant Resistance to Maize Insects, CIMMYT, Mexico DF. 9-14 March 1987:122-129

[21] Raspudić E, Ivezić M, Brmež Majić I. Susceptibility of Croatian maize hybrids to European corn borer. Cereal Research Communications. 1999;**37**:177-180

[22] Bažok R, Igrc Barčić J, Kos T, Gotlin Čuljak T, Šilović M, Jelovčan S, Kozina A. Monitoring and efficacy of selected insecticides for European corn borer (Ostrinia nubilalis Hubn., Lepidoptera: (Crambidae) control. Journal of Pest Science. 2009;**82**(3):311-319

[23] Čamprag D. Euxoa temera, Scotia ypsilon, Scotia segetum. In Čamprag D editor. Priručnik izvještajne i prognozne službe zaštite poljoprivrednih kultura/ Savez društava za zaštitu bilja Jugoslavije, Beograd; 1983. p. 143-148

[24] Novak I. Critical number of *Autographa gamma* L. caterpillars (Lep., Noctuidae) on sugar-beet. Sbornik UVTI – Ochrana Rostlin. 1975;**11**:295-299

[25] Muresanu F, Ciochia V. Flight dynamics of some Lepidoptera species of sugar beet and possibilities their control (Transylvania—Romania). Zbornik Matice srpske za prirodne nauke/Proc. Natural Science. 2006;**110**:209-220

[26] Kravchenko VD, Müller G. Seasonal and spatial distribution of noctuid moths (Lepidoptera: Noctuidae) in the northern and central Arava Valley, Israel. Israel Journal of Entomology. 2008;**38**:19-34

[27] Vajgand D. Flight dynamic of economically important Lepidoptera in Sombor (Serbia) in 2009 and forecast for 2010. Acta Entomologica Serbica. 2009;**14**(2):175-184

[28] Lemic D, Drmić Z, Bažok R. Population dynamics of noctuid moths and damage forecasting in sugar beet. Agricultural and Forest Entomology. 2016;**18**(2):128-136. DOI: 10.1111/afe.12145

[29] Vanparys L. Moth catches of the cabbage moth (*Mamestra brassicae L.*) and the green vegetable noctuid (*Lacanobia oleracea L.*) in West Flanders. Mededeling – Provinciaal Onderzoek- en Voorlichtingscentrum voor Land- en Tuinbouw. Beitem-Roeselare; 1994: 1-4

[30] Giron-Perez K, Nakano O, Silva AC, Oda-Souza M. Attraction of Sphenophorus levis Vaurie adults (Coleoptera: Curculionidae) to vegetal tissues at different conservation level. Neotropical Entomology. 2009;**38**:842-846. DOI: 10.1590/S1519-566X2009000600019

[31] Cohnstaedt LW, Rochon K, Duehl AJ, Anderson JF, Barrera R, Su NY, Gerry AC, Obenauer PJ, Campbell JF, Lysyk TJ, Allan SA. Arthropod surveillance programs: Basic components, strategies and analysis. Annals of Entomological Society of America. 2012; **105**:135-149. DOI: 10.1603/AN11127

[32] Phillips TW, Cogan PM, Fadamiro HY. Pheromones. In: Subramanyam B, Hagstrum DW, editors. Alternatives to Pesticides in Stored-Product IPM. Boston, Massachusetts: Kluwer Academic; 2000. pp. 273-302

[33] Cartea ME, Padilla G, Vilar M, Velasco P. Incidence of the major Brassica pests in Northwestern Spain. Journal of Economic Entomology. 2009;**102**:767-773. DOI: 10.1603/029. 102.0238

[34] Cartea ME, Francisco M, Lema M, Soengas P, Velasco P. Resistance of cabbage (*Brassica oleracea capitata* group) crops to *Mamestra brassicae*. Journal of Economic Entomology. 2010;**103**:1866-1874. DOI: 10.1603/EC09375

[35] Sekulić R, Kereši T. Da li treba hemijski suzbijati repinog moljca? Naučni institut za ratarstvo i povrtlarstvo. Novi Sad, Serbia: Zbornik radova; 2003. p. 38

[36] Fajt E. Repin moljac (*Phthorimaea ocelatela*). Biljna proizvodnja. 1951;**4**(1):136-141

[37] ARM 9® GDM Software. Gylling Data Management Inc. Revision 2018.2, February 6 (B =18046), Brookings, South Dakota. 2015

[38] Townsend GR, Heuberger JV. Methods for estimating losses caused by diseases in fungicide experiments. Plant Disease Report. 1943;**24**:340-343

[39] Kraljević Župić I. Biologija kukuruznog moljca u Sinjskom polju, uz mogućnost biološkog suzbijanja osicom *Trichogramma evanescens* West. (Hym. Trichogrammatidae) [thesis]. Sveučilište u Zagrebu Agronommski fakultet: Zagreb; 1993

[40] Fadamiro HY, Cosst AA, Baker TC. Mating disruption of European corn borer, *Ostrinia nubilalis* by using two types of sex pheromone dispensers deployed in grassy aggregation sites in Iowa cornfields. Journal of Asia-Pacific Entomology. 1999;**2**(2):121-132. DOI: 10.1016/S1226-8615(08)60040-0

[41] Augustinović Z, Raspudić E, Ivezić M, Brmež M, Andreata-Koren M, Ivanek-Martinčić M, Samobor V. Influence of European corn borer (*Ostrinia nubilalis* Hübner) on corn hybrids in north-west and eastern Croatia. Poljoprivreda. 2005;**11**(2):24-29

[42] Raspudić E, Ivezić M, Brmež M, Majić I, Sarajlić A. Intensity of European corn borer (*Ostrinia nubilalis* Hübner) attack in maize monoculture and rotation systems. In: 45. hrvatski i 5. Međunarodni simpozij agronoma, 15-19 veljače 2010; Opatija. Hrvatska. Zbornik Radova. 2010. pp. 901-905

[43] Hegyi Z, Pók I, Szőke C, Pintér J. Chemical quality parameters of maize hybrids in various FAO maturity groups as correlated with yield and yield components. Acta Agronomica Hungarica. 2007;**55**(2):217-225. DOI: 10.1556/AAgr.55.2007.2.9

[44] Djurovic D, Madic M, Bokan N, Stevovic V, Tomic D, Tanaskovic S. Stability parameters for grain yield and its component traits in maize hybrids of different FAO maturity groups. Journal of Central European Agriculture. 2014;**15**(4):199-212

[45] Vulchinkov S, Ilchovska D, Pavlovska B, Ivanova K. Trends in productive abilities of maize hybrids from different FAO groups. Bulgarian Journal of Agricultural Science. 2013;**19**(4):744-749

[46] Vasilj Đ. Biometrika i eksperimentiranje u bilinogojstvu. Hrvatsko agronomsko društvo. Zagreb; 2010

[47] Maceljski M, Balarin I. Faktori dinamike populacije sovice game (*Autographa gamma* L.) u Jugoslaviji. Acta Entomologica Jugoslavica. 1974;**10**:63-76

www.ingramcontent.com/pod-product-compliance
Lightning Source LLC
Chambersburg PA
CBHW081237190326
41458CB00016B/5818